Ten Materials That Shaped Our World

M. Grant Norton

Ten Materials That Shaped Our World

 Springer

Copernicus Books is a brand of Springer

M. Grant Norton
The Honors College
Washington State University
Pullman, WA, USA

ISBN 978-3-030-75212-5   ISBN 978-3-030-75213-2   (eBook)
https://doi.org/10.1007/978-3-030-75213-2

Copernicus is part of Springer, an imprint published by Springer Nature The registered company is Springer
Nature Switzerland AG
The registered company address is: Gewerbestrasse 11, 6330 Cham, Switzerland

Acknowledgments

My aptitude for science would not have matched my interest without the patient guidance of Mr. Roy. He truly was a gifted teacher and, without his hours of gentle encouragement and consistency, I would have fallen at the first hurdle. Dr. Bloss at Godalming Sixth Form College spent as much time teaching A-level Chemistry as he did talking about his Ph.D. research topic: the former was essential, the latter was inspirational. Dr. Brian Cleaver the Undergraduate Admissions Tutor at Southampton University gave me a chance—to avoid me making "a mistake." Dr. John Olby at Cookson Group never lost his enthusiasm for trying to unearth the secrets of manganese pink. I don't think he ever got there, but he approached each trial with optimism and a firm conviction that it was going to be "this time". Quentin Reynolds and Graham Jones were each very supportive. I am not sure that I really made the most of all the opportunities, but visits to see how things are actually made, made me realize the pivotal role that materials play and that I wanted to play a bigger part. The late Prof. Brian C. H. Steele largely left me to my own devices but made sure that it all came together at the end.

My biggest thank you goes to Prof. C. Barry Carter. The Cornell group was amazing and "taught me all I know." The continued collaboration with Barry, that has spanned over 30 years, has been the most rewarding aspect of my professional career. It would be impossible to put that gratitude into appropriate words.

This book, like many, has taken far longer to write than it should have. But it would have taken even longer without the support of the "writing group" at Washington State University. In particular, I would like to thank Peter Chilson, Debbie Lee, Rita Rud, and Annie Lampman. Their encouragement and the enthusiasm with which they greeted each new revision kept me going. I am grateful to my Developmental Editor Amberle Sherman for all her thoughtful insights and probing questions.

My friend and colleague Prof. Su Ha has been, and continues to be, a wonderful collaborator. I think we have made some very good discoveries, which I hope will have a part to play in creating a sustainable future for my son, his boys, and sons and daughters around the world.

I am sorry Ted that your picture doesn't appear in this book, but you make me realize why we need to keep working on finding solutions to all the grand challenges we face. I know that you will do your best.

Finally, Christine, thank you for providing abundant supplies of glassware whether I needed it or not, you know the rest and I am very grateful.

Contents

1	**Introduction**	1
	1.1 Looking at the World as a Materials Scientist	3
	1.2 Why This Book	4
	1.3 Why These Materials	5
	References	6
2	**Flint—*The Material of Evolution***	7
	References	22
3	**Clay—*The Material of Life***	25
	References	44
4	**Iron—*The Material of Industry***	45
	References	63
5	**Gold—*The Material of Empire***	65
	References	84
6	**Glass—*The Material of Clarity***	87
	References	104
7	**Cement—*The Material of Grandeur***	107
	References	123
8	**Rubber—*The Material of Possibilities***	125
	References	143

9   **Polyethylene**—*The Material of Chance*   145
     References   158

10   **Aluminum**—*The Material of Flight*   161
     References   175

11   **Silicon**—*The Material of Information*   177
     References   195

12   **Conclusion**   197
     References   204

Index   207

1

Introduction

Possibly the first time that we looked critically at the world through our changing relationship with materials was when Danish archaeologist and Curator of the National Museum of Denmark Christian Jürgensen Thomsen examined a collection of Scandinavian antiquities and decided to arrange them, not in terms of their shape or properties or function or where they were from, but on the primary material from which they were made. Thomsen's classification produced three distinct groupings of material: stone, bronze, and iron. These became the basis for the popular Three Age system—the Stone Age, the Bronze Age, and the Iron Age—that was published by Thomsen in 1836 and is still widely used by museums today.

Over time, it became clear that there was complexity and subtlety within each of the classifications, which led to further subdivisions. In 1865 English naturalist John Lubbock distinguished the earliest Stone Age period that he called the Paleolithic characterized by flaked flint tools and the much more recent Neolithic where our ancestors worked clay into pottery. Chemically flint and clay have a number of similarities, for instance their main constituents are the elements silicon and oxygen. As materials, clay is very distinct from flint requiring a different understanding of material behavior to shape it in useful objects.

The enormity of the Paleolithic period and the technological innovations that happened over the approximately 2 million years led French prehistorian Gabriel de Mortillet to further divide it into: Lower, Middle, and Upper. Even further subdivision has been used to separate the earliest pebble and flake

© The Author(s), under exclusive license to Springer Nature Switzerland AG 2021
M. G. Norton, *Ten Materials That Shaped Our World*,
https://doi.org/10.1007/978-3-030-75213-2_1

tools discovered in Africa with the appearance of the handaxe, which is associated with two extinct hominin species *Homo erectus* and *Homo heidelbergensis* [1].

The Stone Age produced four important materials technologies. The first was the ability to shape flint by removing flakes to produce tools and weapons. The second was the idea of joining different materials together to increase functionality, when a stone tip was attached to a wooden shaft to form an arrow or a spear or a sickle. Thirdly, the concept of creating an object additively rather than by subtraction—building a pot by adding layers of clay rather than chipping away flakes of flint. And fourth, the importance of fire. All the materials described in this book after flint, and many others, require heat at some stage during their synthesis and processing.

As we learned more about the evolution and spread of bronze and iron technology these metal ages have also been subdivided, although each covers much shorter time periods than the Stone Age. For metals the subdivisions focus less on the material, but more on where and when the technology was being used. Civilizations in Greece began working with bronze before 3000 BCE. In the British Isles the use of bronze began around 1900 BCE and in China even later around 1600 BCE. One of the reasons for this difference, spanning more than 1,000 years, was that for a society to enter into the Bronze Age it required a nearby source of the raw materials; copper and tin. Both are regionally abundant, but neither were widely available to our ancestors without the establishment of robust trading routes. Despite the worldwide availability of flint there is evidence that the most advanced early technologies and societies developed where the highest quality flints were available [2]. So, regional advantages possibly existed prior to the Bronze Age.

The discovery of bronze brought an end to the Stone Age (although not to the end of our use of flint). Bronze in turn gave way to iron. Then many of the applications that would have been satisfied with iron instead used the far superior and more widely useful iron alloy, steel.

Although there has been no formal extension of the Three Age system into a fourth (or more) age, an argument can be made that in terms of identifying a single material that defines our present world more than any other a case could be made that about sixty years ago we entered the Silicon Age. From the first period of human prehistory to the present day we have gone from the Stone Age characterized by flint tools that gave our ancestors an evolutionary advantage to the Silicon Age that enabled social media, artificial intelligence (AI), the Internet of Things (IoT), and has connected almost everyone on the planet.

This book begins with flint, concludes with silicon and in between looks at eight other transformative materials.

1.1 Looking at the World as a Materials Scientist

Using the Three Age system, we can see that materials have an intimate connection with our earliest history. The materials ages cover by far the longest period of our existence; millions of years rather than just the few thousand years from the end of the Iron Age to the present day. The subdivision of these ages has been used to mark important technological changes in our ability to work with the natural world—for instance by shaping flint—and to go beyond the bounds of what nature provides—by combining copper and tin to produce bronze.

Despite our long association with materials, materials science as a discipline only began in the early 1950s. The first university department including the term "materials science" in its name was at Northwestern University in Illinois. The *Journal of Materials Science* established to publish the latest research in the field was created in 1965 and recently celebrated its 1,000th issue [3]. But our study, our examination, of materials goes back to when our ancestors first looked at the sharp curved surfaces of a piece of fractured flint or obsidian and realized it could be used to cut.

When a materials scientist looks at an object, for instance, a Stone Age handaxe the first consideration is its *structure*—a teardrop shape, uneven, but smooth with many conchoidal impressions. Then, its *properties*—the edges are sharp, it is hard, it will abrade wood and scratch metal. *Processing* was required to transform what was once an unassuming and unremarkable pebble into this purposed tool. This transformation was deliberate. It required intent. It required skill. Finally, what was the *performance* of the tool when it was put to its task. How well did it do its job? The field of material science is defined by the interrelationships between structure, properties, processing, and performance, which are typically represented as the four corners of a tetrahedron [4].

This book is very much written from the perspective of a materials scientist. With that context in mind I have attempted to add the why, rather than just the how, certain materials have had the impact they have. For instance, it is the fracture behavior of flint—a direct result of its microstructure, consisting of tiny quartz crystals, formed over millions of years that gave our ancestors the evolutionary advantage of being able to add meat to their diet. When Sir Francis Drake was relieving the Portuguese and Spanish of their gold, he was unaware that the material he sought held its power over Queen Elizabeth I because of the relationship between the outermost electron and the nucleus of the gold atom. But it is that relationship that made gold so desirable for its color and its inertness.

Over time our view of gold has changed. Sir Thomas More, counselor to Henry VIII, saw gold as being "in itself so useless", but it became an essential material—in the form of whisker thin wires—for the fabrication of silicon chips. It is the crystal structure of gold that allows one ounce of the metal to be drawn into a wire 50 miles long. Now 500 years after Sir Thomas, gold is the workhorse of nanotechnology with applications spanning from low emission automobile exhaust catalysts to treating cancer through the delivery of drugs directly to the site of the tumor. It is certainly not useless!

1.2   Why This Book

In this book, I have selected ten materials that have undeniably shaped our world. If these ten materials had not been discovered—or didn't exist—the world as we know it would be very different. There are several books that have been written that take a similar approach to that used here where a materials science professor describes the critical role that materials have played since our earliest ancestors first found or made an object that could be used as a tool. Maybe it happened as imagined by Cornell University professor Stephen Sass where a lump of obsidian was thrown against a rock causing it to shatter into razor-sharp shards that were found to be useful for cutting [5]. Maybe our ancestors found that certain stones were shaped in such a way that they were suited to a specific task; cutting, chopping, scraping. Eventually—slowly— the idea emerged that these stones could be deliberately and carefully shaped to produce a more useful engineered tool.

Although some of the stories associated with these ten materials have been told by others the field is evolving such that there are constantly new discoveries and developments that build on what has already been documented. This is especially true with nanomaterials. For instance, not only has nanoparticle gold challenged our view of this traditional material, but nanomaterials including carbon nanotubes and nanoparticles of silica are being combined with concrete to make it even more durable and stronger [6].

Another example of where we have to update our existing view of a material is glass. We constantly look through glass without even noticing it, unless of course it is dirty or covered in greasy fingerprints, but nanostructured forms of glass are opening up new possibilities for this ubiquitous and ancient material. For instance, tiny glass springs, called nanosprings, have been shown to be effective in trapping exosomes, tiny vesicles excreted by normal and cancerous cells that provide information about the progression of the disease and can possibility help identify the best ways to treat it [7]. This book describes some of these exciting innovations that could impact our future as stone, bronze, and iron impacted the past.

The audience for this book is primarily those that want to learn more about materials and how they affect who we are and how we live our lives. Although not a textbook, the content has been used in a general education course in the sciences taught within the Honors College at Washington State University, a summer course for engineering students at the Chien-Shiung Wu Honors College at Southeast University in Nanjing, and in lectures at Tecnológico De Monterrey at both the Querétaro and San Luis Petosí campuses.

1.3   Why These Materials

The materials described in this book have shaped our world in both large and small ways. Frequently we identify uses that have benefited society, but it is also possible to find instances where our use or quest for materials has been damaging and destructive. The selection of which ten to write about has included some personal bias, which is the prerogative of any author. But the ten do include at least one from each of the primary categories of material: metals, ceramics, polymers, and semiconductors. The materials that were left out suggest possibilities for a future edition.

Diamond—*The Material of Eternity*, which with its superlative hardness is essential for machining everything from lightweight aluminum alloys to high strength concrete and silicon. Since the 15th century diamond has symbolized commitment and although diamonds don't last forever as Shirley Bassey might suggest when she sings the theme tune to the seventh James Bond movie, we are unlikely to witness any spontaneously changing into graphite.

Other contenders might include: Uranium—*The Material of Energy*, the main fuel for nuclear reactors; Plutonium—*The Material of Fear*, one of our synthetic elements that formed the core of the atomic bomb dropped on Nagasaki, Japan; or Graphene—*The Material of Expectation*. Graphene, a sheet structure comprising just a single layer of carbon atoms, has not had an impact equaling that of the ten materials highlighted in this book, but many people think that with its incredible range of properties that it just might.[1]

Notes

1. 11 ways graphene could change the world, https://www.mnn.com/green-tech/research-innovations/stories/10-ways-graphene-could-change-the-world Downloaded January 25, 2019.

References

1. Corbey, R., Jagich, A., Vaesen, K., & Collard, M. (2016). The acheulean handaxe: More like a bird's song than a beatles' tune? *Evolutionary Anthropology, 25,* 6–19.
2. Jacobs, J. (1969). *The Economy of Cities.* New York: Random House.
3. Carter, C. Barry, Norton, M. Grant, & Blanford, Christopher F. (2020). Celebrating 1000 issues. *Journal of Materials Science, 55,* 10281–10283.
4. *Materials Science and Engineering for the 1990's* (1989). Report of the Committee on Materials Science and Engineering, National Research Council, Washington DC: National Academy Press
5. Sass, S. L. (1998). *The Substance of Civilization* (p. 14). New York: Arcade Publishing.
6. Nanosilica refers to nanoparticles of silica glass. A recent paper that describes the benefits of using of nanosilica in cement is: Liu, R., Xiao, H., Liu, J., Guo, S., & Pei, Y. (2019). Improving the microstructure of ITZ and reducing the permeability of concrete with various water/cement rations using nano-silica. *Journal of Materials Science, 54,* 444–456.
7. Ziaei, P., Geruntho, J. J., Marin-Flores, O. G., Berkman, C. E., & Norton, M. G. (2017). Silica nanostructured platform for affinity capture of tumor-derived exosomes. *Journal of Materials Science, 52,* 6907–6916. This paper shows how exosomes from prostate cancer cells can be selectively captured by nanosprings of silica glass that have been "baited" with the appropriate receptor molecules. Exosomes from cancer cells are also referred to as oncosomes: Stone, L. (2017). Sending a signal through oncosomes. *Nature Reviews Urology, 14,* 259.

2

Flint—*The Material of Evolution*

Our information is processed and delivered by tiny silicon chips. Telephone calls and internet data pass at the speed of light under the Atlantic Ocean (and soon the Arctic Ocean) along glass optical fibers that stretch for thousands of miles [1]. We fly around the world in airplanes made of tough aluminum alloys and lightweight carbon-fiber composites, and we live on "platinum" and "gold" credit cards. But two and a half million years ago one material ruled: flint. To our ancient ancestors, flint was an invaluable material because it could be found almost anywhere and, with only a little effort and a lot of patience, a smooth pebble could be transformed into a tool with razor sharp, wear resistant edges. Up until just a few thousand years ago, stone tools made of flint were still widely used for cutting through the hides of animals and butchering their carcasses for food, working and shaping wood that was used to build shelters, and even cracking nuts, a valuable protein-rich snack.

Visiting museums such as the Natural History Museum in London or the American Museum of Natural History in New York City and looking at collections of these early stone tools through eyes acquainted with iPhones, "Dreamliners," Xbox 360s, and all the products of modern technology, they can seem somewhat modest, unimpressive, and certainly primitive, but their importance in our evolutionary process cannot be overestimated. Simply, we probably owe our very existence to the brittleness of flint and the complex way in which it breaks. We are here right now because our ancestors discovered how to transform a piece of flint into a useful tool. As we learned to shape flint, so flint, in turn, has shaped us.

© The Author(s), under exclusive license to Springer Nature Switzerland AG 2021
M. G. Norton, *Ten Materials That Shaped Our World*, https://doi.org/10.1007/978-3-030-75213-2_2

Fig. 2.1 Scanning electron microscope image of a diatom structure. There are a great many variations in these structures, which can only be seen using powerful microscopes. (The image was recorded by J.L. Riesterer and C.B. Carter and originally published in C.B. Carter and M.G. Norton, Ceramic Materials: Science and Engineering, 2nd edition (New York: Springer, 2013),p. 405. Republished here under Springer Copyright Transfer Statement.)

Flint is a naturally occurring sedimentary rock consisting of densely packed quartz crystals that are so small they can only be seen using a microscope. Although it is uncertain how flint was formed, the chemical constituents of flint—silicon and oxygen—are usually accepted to be biogenic, originating from the skeletons of marine organisms such as radiolara and diatoms, which form a silica gel. Figure 2.1 is an electron microscope image of just one variety of the many thousands of species of diatoms. The silica skeleton is an example of a naturally occurring glass, a topic we will meet again in Chap. 6.

Over time as the moist gel dried it began to crystallize forming small quartz crystals. Whilst flint is usually black, changes in the chemical conditions, such as the inclusion of colorful metal sulfides and metal oxides during drying, produced a variety of different colors. The semiprecious gemstones onyx and tigereye are very similar to flint. It is the presence of impurities such as the mineral crocidolite (a blue form of asbestos) and dark brownish red iron oxide that give rise to the colored bands in these pretty stones.

Flint is one of a relatively few types of rock that when broken form sharp hardwearing edges. When flint is chipped the propagating crack twists and turns to follow the boundaries between the densely packed quartz crystals. We describe the fracture as being conchoidal, or shell-like, because the resulting fracture surface resembles the concave shape of a bivalve shell such as a mussel. It is flint's proclivity for conchoidal fracture and its hardness that made it the perfect material for making tools.

Many of the other naturally occurring minerals that were widely available to our ancient ancestors do not undergo conchoidal fracture. Clay and mica, two very abundant silicate minerals, break along well-defined planes of atoms in the crystal structure. These planes are called cleavage planes and coincide with certain crystal faces. In clay and mica, which both have layered structures cleavage occurs between the layers where the atoms are only weakly bonded. It is easiest for a crack to pass *between* adjacent layers rather than propagating *through* a layer where the bonding between the atoms is strong. The familiar soft soapy feel produced when clay is mixed with water is due to cleavage of the clay particles. And, although clay is very useful as we shall see in Chap. 3, it does not have the properties necessary for producing hard-wearing tools for cutting and chopping and would not have provided the evolutionary advantage of flint.

In addition to the way it fractures the other property of flint that makes it particularly suitable for producing cutting and chopping tools is its hardness. At the molecular level the hardness of a material is directly related to the strength of the bonds between constituent atoms. In quartz these atoms are silicon and oxygen. The silicon-oxygen bond is very strong. Flint is a hard mineral because of the hardness of its constituent quartz crystals.

On the hardness scale developed by German mineralogist Fredrich Mohs in 1822, flint has a hardness of 7. To put this number into context, the hardest of all known materials is diamond with hardness on the Mohs' scale of 10. The carbon–carbon bonds in diamond are extremely strong and inflexible. The softest mineral on the scale is talc, which has a hardness of only 1. The basic idea of the Mohs' scale is that a mineral higher on the scale will be able to scratch or abrade one below it. A tool must be harder than the work piece in order to act on it—the harder the material the tool is made out of, the more materials it can work on. So, in addition to preparing food and building shelter, flint tools were used to shape bone (Mohs' hardness 5) into needles to make clothing, and shell (Mohs' hardness 3) into hooks for catching fish.

Shakespeare alludes to the persistence of flint in *Romeo and Juliet*: "Here comes the lady: O! so light a foot will ne'er wear out the everlasting flint." When *Romeo and Juliet* was written—between 1591 and 1595—flint was widely used in both simple and elaborate architectural constructions from cathedrals to farmsteads. Flint buildings define the landscape of many towns and villages in England. It was the material of choice for churches in Norfolk, walls in Hertfordshire, houses in Wiltshire, and barns in Sussex. The Romans built with flint extensively in parts of England where there were abundant supplies that were easy to collect. Flint's hardness and resistance to wind and rain made it an ideal material for fortifications such as castles and city walls. As a young child I would spend the summers with my grandparents in Great Yarmouth. A regular day trip was to visit the Roman fort at Burgh Castle, built in the late third century with flint and brick walls. It is one of the best-preserved Roman monuments in the country.

Flint's hardness and durability, its resistance to weathering, are two of the reasons that stone artifacts have survived over so many years and provide a rich evidence of the prehistoric period. We can compare the abundance of stone tools dating back tens of thousands of years with the comparative lack of Iron Age artifacts lost because of rusting that were just a few thousand years old.

The very earliest flint tools were found in the early 1920s in the Olduvai Gorge in northern Tanzania by Louis and Mary Leakey. The Olduvai Gorge is forty kilometers long and cuts deep into the Serengeti Plain. It is here that anthropologists find the world's best examples of early Stone Age artifacts that are over 2 million years old. The discovery of this archeological site established the great antiquity of human tool making and suggested that Africa (not Asia as some scientists believed at the time) was the cradle of humanity. The current thinking remains that we all descend from African ancestors that migrated out of the continent sometime after 100,000 years ago: Africa is in every one of our DNA [2]. This position on the origin of human evolution was in agreement with that of Charles Darwin, who in 1859 had published the groundbreaking book *On the Origin of Species*. What was also significant about the Leakeys's discoveries was that they pushed back by almost two million years the known dates for the existence of hominin species.

Sir David Attenborough, the famous naturalist and broadcaster, describes his feelings on holding one of the stones brought back from Olduvai Gorge:

Holding this, I can feel what it was like to be out on the African savannahs, needing to cut flesh, for example, to cut into a carcass, in order to get a meal. Picking it up, your first reaction is it's very heavy, and if it's heavy of course it gives power behind your blow. The second is that it fits without any compromise into the palm of the hand, and in a position where there is a sharp edge running from my forefinger to my wrist. So I have in my hand now a sharp knife. And what is more, it's got a bulge on it so I can get a firm grip on the edge, which has been chipped specially and is sharp . . . I could perfectly effectively cut meat with this. That's the sensation I have that links me with the man who actually laboriously chipped it once, twice, three times, four times, five times on one side and three times on the other . . . so eight specific actions by him, knocking it with another stone to take off a flake, and to leave this almost straight line, which is a sharp edge [3].

In most cases, these examples of what was the most advanced technology of their time lay strewn, unrecognizable, among other rocks and pebbles. Chopping tools from Olduvai Gorge have a smooth, rounded base and an irregular, undulating work face, where a few crude flakes have been removed, maybe by simply throwing the stone at a large rock. What made these tools stand out as examples of Stone Age technology to the Leakeys was that they were often found clustered together in groups or lying alongside fragments of bones from animals such as giraffes, antelopes, and elephants. There also is something deliberate about their shape. It does not have the randomness that would be expected from natural processes of weathering and abrasion or having been modified by repeated use. They were manufactured with a clear purpose.

Over time, as our ancestors developed more complex brains and became increasingly dexterous, they created more elaborate flint tools, such as handaxes. With its very distinctive chiseled teardrop shape, the handaxe is regarded as the hallmark of *Homo erectus* ("upright man"), the ancestor of the first *Homo sapiens*. These tools are very different from, and more recognizable to the modern eye, than the earliest artifacts found at Olduvai Gorge. Handaxes are shaped by a process called percussion flaking, which involves chipping away small flakes from one large stone (called the core) by striking it with another stone (called the hammer). Cracks, which form from the impact of the hammer hitting the core, travel at speeds up to 1,000 m per second. That is about three times the velocity of a bullet fired from a 9-mm handgun. The resultant crack edges that form on both the flake and the core are very sharp.

As more and more flakes were removed from the core, the shape of the tools evolved displaying a stronger sense of symmetry. They became increasingly refined demonstrating a greater technological finesse. Mousterian flake tools, named after an archeological site in the Dordogne region of France, are very long and narrow and were made about 50,000 years ago. These flake tools, with their well developed, almost intricate, retouched edges and sharp points, are the technology of the Neanderthal. Mousterian-style tools are found all over Europe, from France to Greece and from England to Hungary, and represent the pinnacle of flint technology.

Through the Stone Age from the earliest tools found at Olduvai Gorge to the Mousterian flake tools there is increasing efficiency in the creation of the cutting edge. Douglas Price, emeritus professor of archeology at the University of Wisconsin-Madison, and Gary Feinman, archeologist at the Field Museum in Chicago, provide an interesting illustration of this increased efficiency: From a 0.5 kg flint a pebble tool would yield 8 cm of cutting

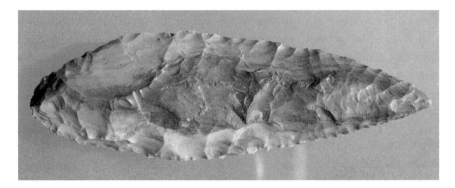

Fig. 2.2 A "high-tech" flint tool. This particular example was found in the state of Washington and donated by the Museum of Archeology at Washington State University. From the fineness of the features it can be determined that the final shape was made by pressure flaking. (Originally published in C.B. Carter and M.G. Norton, Ceramic Materials: Science and Engineering, 2nd edition (New York: Springer, 2013), p. 18. Republished here under Springer Copyright Transfer Statement.)

edge. From a same size flint *Homo erectus* could fashion a handaxe that had a cutting edge of about 30 cm, whereas a flake tool provided about 90 cm of cutting edge, more than ten times that of a pebble tool [4].

There is evidence in the archeological record that some of these refined flake tools were shaped specifically so that they could be attached to handles or shafts. This innovation led to a new generation of more advanced tools and weapons such as handled axes and spears. As early as 200,000 years ago hunters in Africa, the Middle East, and Europe had begun hafting stone points onto the ends of their spears [5]. This was a very important technological step. It was the first time that hominins—our ancestors—had shifted from forming tools by chipping away at rocks to creating tools by joining two different materials into a single tool.

Figure 2.2 shows an example of a flint tool found in the state of Washington in the United States, not far from where I am writing. Although the exact provenance and age of this tool are not known, the basic shape was formed by percussion flaking and then retouched with extra precision using pressure flaking to create the final object. Pressure flaking involves pressing on the core with something sharp, rather than striking it with a stone as in percussion flaking. Moose or deer antler tines and bones were commonly used as instruments for pressure flaking. A skilled flintknapper could precisely control the direction and amount of force that had to be applied to remove very small flakes and produce tools that were sharper and even more intricate that those made simply by percussion flaking alone.

Anthropologists have assigned many applications to flint tools based on their size, shape, and the characteristics of the surface finish. Although it can be difficult to find direct evidence for many of these uses—it is actually pretty much impossible—indirect evidence has come by studying the technologies used by more modern hunter-gatherers such as the !Kung San of the Kalahari Desert, the Pygmies of the Congo basin, and the Aboriginal tribes of Australia. As recently as the 1950s, !Kung San living in an area close to the border between Namibia and Botswana were hunting large game using poisoned-tipped arrows and gathering food in the way that goes back thousands of years. Over the following two decades the traditional hunting grounds made way for game and nature reserves and the !Kung lost large swathes of their traditional hunting lands leaving them little left to hunt and gather.

One of the most important, possibly the defining evolutionary use, of flint tools was hunting for meat and processing the animal carcasses. These tools were a developed advantage that allowed our ancestors to kill and butcher animals despite their comparative lack of physical traits compared with predators in nature. The sharp edges produced by percussion flaking could cut through thick tough animal skins providing access to the protein-rich muscle tissue. Flint tools would have speeded up the time necessary to butcher an animal carcass. The heavy bones and skin could be discarded leaving just the valuable nutritious meat, which could easily be transported. Anthropology professor Kathy Schick, codirector of the Stone Age Institute at Indiana University, has demonstrated how very simple flake tools could cut through the one-inch-thick hide of an African elephant. There is a picture of this procedure in the excellent book by Kathy and her co-author Nicholas Toth, *Making Silent Stones Speak: Human Evolution and the Dawn of Technology* [6].

Many studies have found correlations between meat intake, fertility, intelligence, good health, and longevity [7, 8]. According to University of California-Berkeley anthropologist Katharine Milton, a new nutrient-rich meat diet enabled by stone tools provided the catalyst for human evolution, particularly the growth of the brain. "I have come to believe that the incorporation of animal matter into the diet played an absolutely essential role in human evolution" [9]. Eating meat, which supplies essential amino acids, at a young age would have helped children's brains to grow and develop more quickly. By including meat in their diet our human ancestors became smarter, bigger, and stronger, which ultimately led to their evolutionary success particularly as they spread out across Africa and into Asia.

Stone tools not only enabled an evolutionary change they also laid the foundations for human civilization. The sharp edges could be used to prepare the ground for planting and subsequent harvesting of crops, which led to an enormous societal change from hunting animals and collecting wild plants to a society based on farming for food production. Farming was certainly in evidence as long ago as ten thousand years in Southwest Asia. Later, ancient Egyptian tomb paintings, such as those in the tomb of Sennedjem in Thebes (near Luxor) dating from 1200 BCE, illustrate harvesting of cereal crops using flint sickle blades with wooden or bone handles.

Stone tools played a critical role not only in the establishment of farming but also in its expansion. After the end of the Ice Age, about 12,000 years ago, dense forests began to spread over the tundra and steppe. These forests covered land necessary for growing crops and Stone Age farmers set about clearing these forests with great abandon using their flint-headed axes. A number of studies, mostly conducted in Denmark, have shown how effective flint axe heads were for clearing forests [10]. In one particular study, it took three men just 4 h to clear 600 square yards of silver birch forest. More than 100 trees alone were felled with one axe-head, which had not been sharpened for about 4,000 years! This result is a remarkable testament to the durability of stone tools.

As demand increased and easily accessible surface sources of flint became depleted it was necessary to go underground. A cottage industry developed around the mining, processing, and trading of flint. Flints were mined by digging a large hole down to the flint "floorstone," which would be several meters below ground level. Radiating out from the floorstone were the seams containing the prized flint. For picks, miners used deer antlers, which could be sharpened and then driven into the seams. The antlers were just the right shape for levering out the flint. For shovels, miners used deer shoulder blades from which the spine had been trimmed to give the maximum amount of flat area. A historically important Stone Age flint mine that is open to visitors is Grime's Graves, near Thetford in Norfolk in the UK.[1] Here the prehistoric miners were after the best quality jet-black flint that could be easily "knapped" or shaped for axe heads. Artifacts found at this site provide insights into how flint was mined. The pockmarked site that opened to the public in Spring 2017 contains over 400 shafts, pits, quarries, and spoil dumps. Modern day visitors to the English Heritage site descend 9 m (about 30 feet) by ladder into one of the excavated shafts, which are at a year-round temperature of 50 °F. The name "Grime's Graves" has an Anglo-Saxon origin and means "the Devil's holes."

As processes to produce metals, particularly bronze beginning around 3000 BCE and then iron from about 500 BCE, became developed and disseminated, the technological importance of flint declined in many parts of the world. This decline was particularly evident in the Near East, Europe, and China. At the dawn of the Bronze Age, about 5,000 years ago, a technology based on metals began to emerge. By combining the metallic elements copper and tin it was possible for metal tools with properties far superior to flint to be manufactured. Bronze was first in common use in the eastern Mediterranean countries, mainly Crete, Greece, and Turkey. In northern Thailand, where both copper and tin are found, bronze production was well established by 2000 BCE. By the second millennium BCE very large bronze objects could be made such as the enormous 800 kg bronze cauldron cast in China and on display, along with many other early Bronze Age objects, in the National Museum of China located on the eastern side of Tiananmen Square in Beijing.

As is the case with the development of many modern materials, one of the first applications of bronze was military. It was initially used to form simple flat daggers, but the defining use was in swords. The ease with which bronze could be melted, cast and then hammered into complex shapes ushered in the age of the sword.

Bronze also provided some significant improvements for domestic applications such as farming. Metal plows made of bronze were far superior to those made from wood and stone. The increasing amounts of land that could be cultivated using bronze plows enabled the large populations of the early Mesopotamian cities of Ur, Urak, and Eridu to be fed. Evidence for this change in agricultural practice is frequently provided from indirect sources such as the figurines of yoked oxen that were found at Tsoungiza Hill, an early Bronze Age site in the Peloponnese region of southern Greece.

Not only plows were improved by the use of bronze. Bronze bits or cheek pieces made it easier for farmers and drovers to control the horses, asses, and oxen used for pulling plows and carts because leather bits could not survive the incessant grinding of teeth and the corrosive effect of saliva. The combination of better plows and the use of animal power made a tremendous impact on the amount of land that could be farmed and significantly increased productivity. Although fashioning bronze tools and implements would have been slow by today's standards of mass production, bronze tools could be turned out far more quickly than stone tools. With its ease of production and superior performance bronze tools reduced the importance of stone in a number of applications.

But there were still important uses for flint.

The arrival around 2000 BCE of the so-called Beaker people (who most likely migrated northward from southern Spain) marked the beginning of the Bronze Age in Britain. In addition to their skill with bronze, these warrior invaders had a process for making pottery, which they used to create beakers, urns, and other forms of container. These items were decorated with lines and chevron patterns made using either a cord or a square-toothed comb. The markings provided an aesthetic value and made the containers easier to hold. They acted like grips. But although the Beaker ware, with its lines and chevrons, was elaborate in its decoration, it was quite weak, lacking even the moderate toughness of today's ceramics. It was not able to withstand direct heating in a fire. To cook food the Beaker people devised a clever indirect heating system. Flint nodules were heated in the fire and when they were hot enough they were plunged into the pots containing water and the other ingredients. Because flint can withstand very high temperatures (quartz doesn't melt until almost 1700 °C, over 3000°F), it would have been possible to boil water and cook food in this way. However, the repeated heating and rapid quenching into water eventually cracked the flint. Heaps several yards in diameter and a few feet thick, of cracked flints together with evidence of wood ash, can be found on sites where the Beaker people lived.

At some stage during the Paleolithic period, about one million years ago, when the handaxe was well established and more refined flint tools like the Clactonian chopping tools were being made and used, the ability to use flint for producing sparks and kindling a fire was discovered. The important property of flint that allows it to be used in this application is its hardness. Flint artifacts that were possibly used as strike-a-lights are occasionally found. They are characterized by polished edges caused by being struck repeatedly against crystals of iron pyrite, which would give a shower of sparks sufficient to ignite a small mound of dried leaves and branches. Iron pyrite, the mineral iron sulfide, is abundant and often found with other types of rock in quartz veins. It is commonly known as "fool's gold" because miners—particularly during the gold rushes—often in their excitement mistook it for the real thing.

Using iron pyrite and flint to make fire may have been discovered when sparks were accidentally produced when a piece of iron pyrite was used as a hammer stone during percussion flaking. Although we will likely never know the exact origin of how fire was made, there is no doubt about how important fire was for heating, cooking, and eventually for making metals.

After the discovery of iron, flint was still used as a strike-a-light. In one common, but slightly risky sounding, arrangement a piece of flint was struck against a strip of iron held across the knuckles. In some countries the use of strike-a-lights for lighting candles acquired a religious significance. For example, on Maundy Thursday (the day before Good Friday) it was customary in the Roman Catholic Church to extinguish the candles and to relight them with new fire produced using flint and steel: "At the ninth hour a fire is produced by a flint and steel, sufficient to light a candle, which ought to be placed on a reed; a lamp lighted from this is kept unextinguished in the Church until Easter eve, to light the Paschal taper, which is to be blessed on that day." It was only in 1970 that the Church of Rome permitted the use of a simple but more efficient cigarette lighter for lighting the Paschal candle.

The ability of flint to start a fire eventually was exploited in a new type of deadly weapon, the flintlock. The English Civil War, which lasted from 1642 to 1651, was the first time in the history of warfare that firearms were at least as important as other weapons in influencing the outcome of a battle. The most important surviving group of flintlock muskets from the English Civil War is in the Great Hall of Littlecote House, Wiltshire in the UK.

The flintlock was introduced in the first half of the seventeenth century and was used for over 200 years in military and civilian firearms. It had two very obvious advantages: it was simple to use and it was reliable. Prior to the development of the flintlock, guns had very limited use for warfare. The problems were that muzzle loading was difficult and the matchlock ignition system, which predated the flintlock, was unreliable. The 'match' was a smoldering piece of cord that was easily extinguished in wet or windy weather. Relighting could take precious—in some cases, a fatally long—time.

In 1517, the wheel lock for spark ignition was invented. This mechanism used iron pyrite crystals and steel. A steel wheel with serrated edges, turned by a strong spring, was dragged across the crystal of pyrite. A shower of sparks was created that ignited the priming powder. This ignition system did not suffer in bad weather, but it did have its own set of problems. The mechanism was complicated and difficult to make, so the guns were expensive. Secondly, pyrite is brittle and was prone to breaking. It would then be necessary to readjust the position of the remaining piece of pyrite or replace it altogether. Both processes took time, which might not be a problem for the pheasant hunter but could be fatal on the battlefield.

A more reliable and inexpensive source of sparks could be obtained from flint and steel; as a result, the flintlock was developed. The flint was held in the jaws of the hammer, which could be pulled back and locked into the 'cocked' position. An L-shaped piece of steel called the 'frizzen' covered the priming powder and was held in position by a spring. When the trigger was pulled the hammer was released, the flint struck the frizzen, pushing it back, and a shower of sparks was sent towards the now exposed priming powder. For safety, the cock could be held in a halfway position—'half-cocked'—and the trigger could not then be operated.[2] In 1645, a matchlock could cost as much as eleven shillings and six pence (equivalent to a purchasing power today of about $150); the cheapest flintlock would be fourteen shillings and sixpence. Flintlocks were more expensive because the mechanism, although simple to operate, was more complicated to make.

When Grenadier (French for "Grenade man") units were formed in 1671, the soldiers were issued flintlock carbines—essentially shorter muskets—fitted with slings for carrying them on their shoulders. The slings allowed both arms to be free for throwing grenades. Another name that was used for the carbines was 'fuzil,' which was derived from 'fucile' the Italian word for flint. The word eventually became corrupted into 'fusiliers' by the troops armed with these weapons. By about 1700 most matchlocks had been replaced by flintlocks.

Highwaymen, such as the legendary Dick Turpin, favored the flintlock pistol because it was small and easily concealed. Once the flintlock was loaded and cocked it was ready to fire instantly. The shooter had only to aim and pull the trigger. The English Army used the flintlock mechanism for its legendary "Brown Bess" musket, which became the main battlefield firearm around 1720 and was used for over 120 years. Because of the flintlock's reliability it was also used on naval cannon and on pistols for the cavalry. During the Napoleonic Wars the use of flintlocks peaked: 3.5 million muskets were manufactured for the British Army alone. In 1804, the British Army contracted with gunflint makers in Brandon, a small town in Suffolk for the supply of 360,000 flints per month, priced at twenty-one shillings per thousand. By the end of the Napoleonic Wars, Brandon gunflints—mined just four miles from Grime's Graves—were being exported all around the world: to North and South America, Africa, New Zealand, Spain, Russia, China, and Malaya. During the Crimean War, Brandon supplied eleven million flints a year to the Turkish Army. Brandon flints were still in use in Abyssinia in 1935 as the poorly armed Ethiopians fought against Mussolini's well-equipped Italian army. Even into the middle of the twentieth century, two thousand gunflints were being made each day, mainly for export to countries in Africa.

The flintlock was eventually replaced by percussion ignition, which was invented during the height of the flintlock era in 1806 by the Reverend Alexander John Forsyth, a church minister from Belhelvie in Aberdeenshire, Scotland, and a keen wildfowl hunter.

As metals became used for tools, stone tools were left behind, their original purpose and use were forgotten. Flint tools that were once the height of "high-tech" assumed a magical and cultic significance. In medieval Europe discarded stone tools, which would have been made by some of the most talented Stone Age craftsmen, were carried simply as amulets. In European folklore, handaxes, the hallmark of *Homo erectus*, were known as "thunderstones." There was a common belief at the time that these pointed-shaped stones were projectiles from space carried by bolts of lightning that hit the ground [11]. Flint arrowheads found in early farming communities were sometimes called "elfshot": the points of tiny arrows fired by elves. The true origin of these flint tools was not realized or appreciated until much later. It took time, evidence, and even more argument to confirm the connection between stone tools and human ancestry.

In 1800, John Frere, an English country squire and former high sheriff of the county of Suffolk, published a short account describing some stone implements he had unearthed in the village of Hoxne (pronounced "Hoxon"). Frere described these implements that were found scattered amongst bones of extinct animals (such as elephants, rhinoceros, and lion) as "weapons of war, fabricated and used by a people who had not the use of metals." He concluded that the artifacts must date from "a very remote period indeed; even beyond that of the present world." An example of one of Frere's Hoxne handaxes is on display at the British Museum in London.[3] The classic teardrop shape had been carefully flaked on both faces to produce two keen cutting edges and a very sharp narrow point, which suggest that it might have been used as a general-purpose butchery tool rather than a weapon of war. Hoxne has produced a wealth of ancient archeological objects, including the famous Hoxne hoard, which was unearthed in 1992. The hoard is the richest collection of treasure from Roman Britain and contains over 15,000 gold, silver, and bronze coins among many other precious objects such as silver ladles and gold jewelry. These objects were all buried for safekeeping toward the end of the Roman occupation of Britain.

Surprisingly, Frere's paper, which was published in the journal *Archaeologia*, went unnoticed for more than 50 years [12]. Then almost suddenly there was a collective realization among many archeologists, scientists, and historians that humans had an ancient history and that the artifacts described by Frere were part of antiquity. So, it was during the Victorian era that the connection was made between the stone artifacts that were being found at a number of sites across Europe and our prehistoric past. For example, the Grime's Graves site that dates back over 5,000 years was not excavated until 1870.

These ancient relics that represented a bygone time fascinated a Victorian public that was becoming increasingly interested with many aspects of science, technology, and engineering. The desire to own examples of handaxes, arrowheads, and other stone tools for personal collections created a market for these objects that could be satisfied from both legitimate and nonlegitimate sources. One person more than willing to satisfy the demands of curious and gullible Victorians was master flintknapper Edward Simpson. Better known as "Flint Jack," Simpson was one of the earliest experimental stone toolmakers. He was fascinated with stone tools and became remarkably proficient in making replicas of the flint tools being uncovered at the time in ancient river terraces of the River Thames in London and in other parts of England. These replicas he sold in the late nineteenth century to museums and to an unsuspecting Victorian public.

Simpson normally used nothing more than a steel hammer to fashion his "Stone Age" tools out of flint pebbles. By smearing the shaped artifacts with dirt or soot, which he rubbed in with a cabbage leaf, he was able to make them appear worn and aged. Ancient flint tools often acquire a distinctive patination produced by long-term exposure to water and soil. Minerals in the soil can be absorbed into the flint and produce a range of striking colors from reds to yellows to greens. The color can often be used as a "fingerprint" to pinpoint where the flints were buried. The peaty Fens in Norfolk produces flints that are often dark brown. A speckled patina of yellow and green (known to collectors as 'toad-belly') is peculiar to surface implements found on Warren Hill in Suffolk. Weathering can modify the texture of the flint making it rougher and uneven, which produces a milky-white appearance. The absence of patination is a good indication that there is no great age to the flint tool. Even under accelerated aging conditions it takes about two years to produce even a thin layer of patination.

In 1867 Edward Simpson was sent to prison for theft and it is in prison that he is believed to have died. Simpson's story, at least many of the specifics, is not well documented. However, fine examples of his work can be seen in Reading Museum and in Salisbury Museum, both in the UK [13]. Simpson once coolly boasted to his one-time colleague Professor Tennant, of London, that there were also "plenty of his things in the British Museum—and very good things they were too."

Examples of flint tools covering various periods of the Stone Age, from the earliest Paleolithic through to the Neolithic, are staples in the collections of museums and universities around the world. These artifacts of an earlier time continue to yield information that allows us to understand why they were effective and how they were used. Analyzing stone tools using high-power optical and electron microscopes provides evidence that the combined action of pressure, frictional heat, and water can transform the original crystalline quartz surface into a silica gel. When use of the tool has stopped, the gel solidifies forming a very sharp highly polished glassy surface. Analysis of wear patterns and the chemical composition of the cutting edges also sheds light on the possible applications for the tool and how often it may have been used [14]. More recent studies using a wide range of advanced instrumentation to examine several hundred tiny flint flakes around 2 to 3 cm long that were between 300,000 to 500,000 years old found evidence of recycling. The flakes had been made from larger recycled tools to create the equivalent of a modern-day kitchen knife set [15].

Although flint will remain an abundant resource long after many other materials become depleted, metals now fill all the roles that flint tools played. Despite its unparalleled importance in our evolution as a species and in the establishment of human civilization, flint is no longer technologically significant. Flint tools remain as evidence of human development and are important in understanding the lives our ancestors lived. The dominance of flint over such an enormous period of human history makes it—along with glass, which is also mainly silica—very special among all the materials that have shaped our world. In comparison, most plastics have been around for much less than one hundred years (the eighty-year anniversary of the production of low-density polyethylene was only celebrated in 2019) and silicon was not made into "chips" until the early 1960s.

The history of flint intersects with two of the other materials described in this book: clay and glass. Flint powder was added to clay in the manufacture of pottery. It stiffened the clay mix making it easier to work and enhanced the whiteness of the final product. Italian glassmakers, famous for their Venetian *cristallo* glass, used flint pebbles they obtained from the bed of the river Ticino as a source of silica for glass making.

Notes

1. Grime's Graves is an English Heritage site. www.english-heritage.org.uk/daysout/properties/grimes-graves-prehistoric-flint-mine.
2. Sometimes accidents did occur and the gun was said to go off at 'half cock.' The expression 'going off at half cock' is still used today when something is unexpectedly or inadvertently set in motion and dates back to these early flintlock accidents.
3. This link shows a picture of one of the Hoxne handaxes. Currently, this handaxe can be seen in the Enlightenment gallery at the British Museum in London. www.britishmuseum.org/explore/highlights/highlight_objects/pe_prb/h/hoxne_handaxe.

References

1. Agrawal G. P. (2016). Optical communication: Its history and recent progress. In: Al-Amri M., El-Gomati M., Zubairy M. (eds) *Optics in Our Time*. Cham: Springer.
2. Ingman, M., Kaessmann, H., Pääbo, S., & Gyllensten, U. (2000). Mitochondrial genome variation and the origin of modern humans. *Nature, 408*, 708–712.
3. Quoted in MacGregor, Neil (2012). *A History of the World in 100 Objects* (p. 9). London: Penguin Books.
4. Price, T. D., & Feinman, G. M. (2005). *Images of the Past* (4th ed., p. 76). Boston: McGraw-Hill.
5. Rots, V., & Van Peer, P. (2006). Early evidence of complexity in lithic economy: Core-axe production, hafting and use at Late Middle Pleistocene site 8-B-11, Sai Island (Sudan). *Journal of Archaeological Science, 33*, 360–371.
6. Schick, Kathy & Toth, Nicholas (1993). *Making Silent Stones Speak: Human Evolution and the Dawn of Technology*. New York: Simon & Schuster. This is an excellent book with a detailed description of the development of stone tool technology.
7. Psouni, E., Janke, A., & Garwicz, M. (2012). Impact of carnivory on human development and evolution revealed by a new unifying model of weaning in mammals. *PLoS ONE, 7*, e32452.
8. Williams, A. C., & Hill, L. J. (2017). Meat and nicotinamide: A causal role in human evolution, history, and demographics. *International Journal of Tryptophan Research, 10*, 1178646917704661.
9. Milton, K. (1999). A hypothesis to explain the role of meat-eating in human evolution. *Evolutionary Anthropology, 8*, 11.
10. Cole, Sonia (1963). *The Neolithic Revolution* (3rd edition). London: The British Museum (Natural History). There are pictures of the Danish Stone Age axe experiment on page 61.
11. Lukis, J. W. (1875). On St. Lythan's and St. Nicholas' Cromlechs and other remains near Cardiff. *Archaeologia cambrensis* (p. 177). London: J. Parker.
12. Frere, J. (1800). Flint weapons discovered at Hoxne in Suffolk, *Archaeologia, 13*, 204. *Archaeologia* was published between 1770–1991 and the abstract of Frere's paper states: "I take the liberty to request you to lay before the Society some flints found in the parish of Hoxne, in the county of Suffolk, which, if not particularly objects of curiosity in themselves, must, I think, be considered in that light, from the situation in which they were found."
13. Stevens, Joseph (1894). *Flint Jack: A Short History of A Notorious Forger of Antiquities*. Reading: Reading Museum. A short history of "Flint Jack," aka "Fossil Willy," aka "Old Antiquarian," aka "Cockney Bill," aka "Bones," aka "Shirtless." I am grateful to the staff of the Reading Museum for making me aware of

this pamphlet and providing me a photocopy. The Reading Museum has some of Flint Jack's known handiwork on display.

14. Jensen, H. J. (1988). Functional analysis of prehistoric flint tools by high-power microscopy: A review of west European research. *Journal of World Prehistory, 2,* 53–88.
15. Venditti, Flavia, Cristiani, Emanuela, Nunziante-Cesaro, Stella, Agam, Aviad, Lemorini, Cristina, & Barkai, Ran (2019). Animal residues found on tiny Lower Paleolithic tools reveal their use in butchery. *Scientific Reports, 9,* 13031.

3

Clay—*The Material of Life*

The symbolic significance of clay can be traced back to very early times when ancient mythologies describe man as being created from clay. In Jewish folklore, for example, a golem is an animated being made from red clay (an impure form of clay that is used in making earthenware pots). The name "golem" possibly comes from the Hebrew word *gelem*, which means simply "raw material." Clay is the raw material that is most widely used in making our everyday ceramic items from cups and saucers to toilet bowls and sink basins. When we die, we are said to return to clay or, in the tragic Irish folk song "The Wind That Shakes the Barley" written by Robert Dwyer Joyce about the 1798 rising, on death our bodies become cold like clay.

Clay, then, is a material associated with both life and death. In practice, clay has a similar dual association. On the one hand, clay pots were essential for collecting and carrying life-giving water and storing grain for the people in the world's first emerging cities in what is now war-torn Iraq. On the other, clay urns were—and are still—used as the final resting place for human ashes. Clay figures were buried alongside Egyptian pharaohs because they were considered essential companions for the afterlife. In early Chinese culture it was common for wives and servants to be sacrificed and buried alongside important rulers. The great philosopher Confucius, probably among many others, condemned this barbaric practice and clay models eventually and thankfully, became used instead of real humans. Collections of funerary clay figurines such as those shown in Fig. 3.1 from the Nanjing Museum are common. Miniature clay pigs—an important animal in Chinese culture

© The Author(s), under exclusive license to Springer Nature Switzerland AG 2021
M. G. Norton, *Ten Materials That Shaped Our World*,
https://doi.org/10.1007/978-3-030-75213-2_3

Fig. 3.1 A group of funerary clay figurines on display at the Nanjing Museum

because it is a symbol of good luck and prosperity—have been found in graves from the Han Dynasty (206 BCE to 220 CE).

The custom of burying a variety of ceramic objects with the dead has resulted in the survival of large quantities of well-preserved pottery. In some cultures, for example the Moche, an Andean civilization that flourished in Northern Peru from the 1st to the eighth century, the pots depict scenes of everyday life: from eating and hunting, to lovemaking and punishment. The pottery provides us evidence of a social history that in all likelihood would otherwise have been lost. Pottery yields a rich historical record of date, social status, household routine, and trade. In the United States, the Jamestown Ceramics Research Group is examining clay objects found in pre-1650 Jamestown and the surrounding area as a way of understanding the early seventeenth century history of the region. The hope is that the pottery will provide insights into the activities and diet of the early settlers as well as uncovering the establishment of trade patterns.

Clay has also given us the earliest written descriptions of life. Clay tablets from Mesopotamia made about 5,000 years ago depict a series of symbols (called pictograms), many of which can very easily be identified with the objects they were obviously made to represent such as cups, bowls filled with food, various animals, and people. There are circles of different sizes, which may be celestial bodies such as the moon and sun and those planets bright enough to see with the naked eye. The symbols were sketched onto the surface of tablets of damp clay, dried and stored. Because of the durability of dry clay many tablets have survived in remarkably good condition. Pictograms became replaced by ideograms, which became commonly understood symbols that might show a clear resemblance to the actual object itself. Ideograms then became replaced in a number of places by cuneiform, where reeds with triangular cross section were used to make impressions that looked nothing like any real object but represented the beginning of writing.

Most of the clay artifacts that are found have been fired. Even after it has been fired, clay is very brittle. An archeologist will find very few unbroken pottery artifacts during a dig, yet each of the individual fragments has survived burial in some cases for thousands of years. During this time their internal structure has remained unchanged.

Remnants of ancient pottery production have been unearthed from almost every part of the world. These discoveries show how important and widespread this technology was, which is perhaps not surprising considering that the raw materials necessary for pottery production are the most abundant minerals on earth. Combined, oxygen (O), silicon (Si), and aluminum (Al) account for about 85% of the elements in the Earth's crust.[1] Consequently, the dominant terrestrial minerals are silicates, which are compounds containing silicon and oxygen, such as quartz (chemical formula, SiO_2), and aluminosilicates, which are compounds containing silicon, oxygen and aluminum, such as kaolin, which has the chemical formula $Al_2Si_2O_5(OH)_4$. The abundance and broad distribution of silicates and aluminosilicates are why pottery was so widespread. The ease with which wet clay can be shaped and the low temperatures needed to bake it are why many civilizations were able to independently develop ceramic technology so successfully.

The behavior of clay, like that of most materials, is directly related to its structure.[2] Clay minerals are layered compounds. Structurally, they have a lot in common with graphite, the "lead" in a pencil. The basic structural units in clay are the tetrahedron (a triangular-based pyramid) and the octahedron (two square-based pyramids sharing bases). These shapes are shown in Fig. 3.2. One silicon atom and four oxygen atoms form tiny tetrahedra. The distance between the silicon and oxygen atoms is about 0.16 nm. The tetrahedra join corners to form extended sheets. One aluminum atom, three oxygens, and three hydroxyl groups (OH) form octahedra. These octahedra also share corners with each other to form sheets. The structure of clay resembles a multi-deck sandwich consisting of alternating sheets of linked tetrahedra and octahedra. The clay particles themselves consist of platelets, each containing many of these sheets. Internal stresses formed because of a size mismatch between the tetrahedral sheets and the octahedral sheets limit the size of the clay particles to less than 0.002 mm. When separated by a film

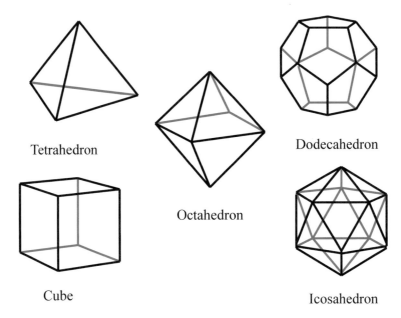

Fig. 3.2 The tetrahedron and the octahedron are the basic building blocks of all the clay minerals. They are two of the five Platonic solids, which were classified in 400 BCE by Plato. The cube is also an important shape shown by a number of different materials such as salt, although we often describe the structure of salt as consisting of edge sharing octahedra

of water, the particles slide over one another allowing a clay-water mixture to be easily shaped by hand.

The very oldest recorded clay objects certainly had a symbolic rather than practical relevance. In 1920, more than ten thousand fragments of statuettes were found near Dolní Vestonice, Moravia, in the eastern part of the Czech Republic. They portray wolves, horses, foxes, birds, cats, bears, and women. One of these prehistoric female figures, shown in Fig. 3.3, remarkably remained almost undamaged. The statuette, known as the "Venus of Vestonice", stands about 10 cm tall and has been dated as far back as 23,000 BCE. The function of "Venus" and many of the other similar Venus-figurines, which have been found stretching across Eurasia from southern France to Siberia, has been widely debated by archaeologists and anthropologists over many decades [1]. They have been described as fertility charms. Alternative interpretations suggest that they were made by women representing a woman's view of their own bodies when viewed top-down. These self-depictions that focus on the obstetrical and gynecological features with the complete absence of any facial characteristics may reflect a woman's reproductive role [2]. We will likely never know the true interpretation of these

Fig. 3.3 The "Venus of Vestonice". This 25,000-year old baked clay figurine was found in 1920 in Dolni Vestonice in the Czech Republic. It is in the Moravske Museum, Brno, Czech Republic. (Originally published in C.B. Carter and M.G. Norton, *Ceramic Materials: Science and Engineering*, 2nd edition (New York: Springer, 2013), p. 19. Republished here under Springer Copyright Transfer Statement.)

Venus figures or even whether they were crafted by women or men, but they provide important evidence of our ancestor's skill in shaping clay.

One of the most wonderful archeological finds, which demonstrated our ancestor's mastery of clay, was in the caves of Tuc d'Audobert in the Hautes-Pyrénées region of France. On October 10, 1912 beautifully preserved clay bison were found that are estimated to be 12,000 years old. These two-foot-high raw clay sculptures of a cow and a bull are captured just before the act of conception as the bull rises above the poised cow. Bison roamed wild in this part of France and would have been an important resource, just as they were for the Native Americans in the American plains. These figures and cave art featuring bison possibly were created to ensure successful continuation of the herds, as they were a major source of both food and clothing.

About 10,000 BCE, coinciding with the establishment of more settled communities, there was a notable shift in the use of clay. From being molded into animal sculptures and figurines this versatile material was used to create functional pottery. Some of the earliest archeological examples of pottery production date to the discovery of fragments from a cave dwelling near Nagasaki, Japan. These fragments from narrow-necked containers ideal for holding water and wide-mouthed vessels suitable for storing grain were produced by the Jomon culture of Japan, a hunter-gatherer society. Jomon pottery is so called because of the characteristic surface patterns made with a twisted cord or rope. ("Jomon" means simply "cord pattern.") Incised lines made with sticks, bones, or fingernails further decorate the vessels with lines, chevrons, and small indents. These pots were fashioned by hand from moist clay, dried in the sun, then baked in a fire to transform the soft fragile shapes into hard durable objects. The warmth of the sun drives off the water that is present between the clay particles causing them to draw closer together. In the heat of the fire the clay particles fuse together forming a hard and durable ceramic in a process called sintering.

It is not clear how our ancestors first made the discovery that heating clay until it was a dull red (this corresponds to a temperature of about 600 °C or 1,112 °F: close to the melting temperature of aluminum) transformed it from a malleable paste into a hardened functional material.[3] One theory I particularly like—as an enthusiastic birdwatcher—is that a bird's nest might have been used as a template and smeared with wet clay [3]. After drying, these simple cups were used to cook food over a fire that baked the clay hard. From this humble beginning pottery makers became increasingly ambitious and their products more sophisticated. And as firing techniques developed the resulting product became more durable and complex.

It would have been quite easy to obtain temperatures of 600 °C or even greater using nothing more than an open bonfire. The dried pots were placed on the ground along with a suitable fuel: animal dung worked quite well as did dried reeds and rushes. This basic approach is still used in parts of South America, Africa, and the Middle East. More elaborate "kilns" (the ceramic industry term for furnace) developed slowly. The key step was understanding and being able to control airflow.

In an updraft kiln the dried pots were stacked on pedestals and covered. Fuel is burnt at the bottom of the kiln. The hot air rises to the region where the pots are placed and there is an opening at the top of the kiln, which acts as a chimney. By maintaining the flow of hot air around the pots they could reach temperatures as high as 1000 °C. The Romans developed some elaborate updraft kilns. Of course, they had an excellent understanding of heat flow because of their work with hot baths. Modifications of the basic design of the updraft kiln such as using a perforated grate to control the flame direction led to the attainment of even higher temperatures and more uniformity of the temperature within the kiln. The kiln technology that was developed for making pottery would become very important a little later in history, because it would be needed for extracting iron from iron ore.

The process of making good pots would have involved a considerable amount of trial and error. Firstly, it was necessary to make the clay wet enough so that it could be shaped. Too wet and it would slump uncontrollably; too dry and it would be sticky and difficult to shape. Once shaped, the damp pot would have to be dried slowly in the sun before it could be baked. In the modern-day mass production of large ceramics such as toilet bowls it is important for the items to be dried very slowly to avoid cracking. Drying cycles can take a whole day and the firing process a couple of days or more.

Early potters found that the addition of straw to the wet clay acted as a binder, making a stronger composite brick. This approach is what we continue to do today with materials such as fiberglass, which is a moldable plastic reinforced with glass fibers. In high tech composites, carbon fibers and more recently graphite nanofibers, carbon nanotubes, and even graphene are being added to plastics to form very lightweight and high strength materials.

From the very beginning, the appearance of pottery was almost as important as its function. It had to look good. Ways to impart colors that were not the natural brownish red of the clay were developed. By covering the bonfire with wet leaves the pots underneath would blacken because carbon monoxide in the smoke reduced oxides, primarily iron oxides, in the clay to their metallic state by removing oxygen. In turn, the process converts carbon monoxide into carbon dioxide. Creating a reducing atmosphere (one that lacks oxygen) within the kiln or furnace was possibly another of those fortunate accidental discoveries made by our early ancestors.

Reduction firing as it is termed has been traced back as far as 500 BCE in China. [4] These early Chinese potters knew the difference between firing in an oxidizing atmosphere (one rich in air) and a reducing atmosphere and the effect that that would have on the fired clay.

It is not only the atmosphere inside the kiln that can change the properties of the ceramic. The kiln temperature also has an important effect. We often divide traditional pottery into two broad classes based on their firing temperature. The classifications—earthenware and stoneware—are not particularly useful other than for museums to arrange their collections, but they do demonstrate a major technological change: our ability to reach higher temperatures.

The earliest form of pottery is earthenware. Earthenware is made from red "earthenware clay" that has been heated to around 1000 °C. All of the pottery that was made more than 3,500 years ago we would class as earthenware. It is porous and often red. The term "pottery" is usually synonymous with earthenware. Because clay occurs naturally over much of the Earth's surface, we find examples of earthenware from many parts of the world. Regional differences in the physical and chemical characteristics of the local clay, and the skill of the potter, gave rise to variations in the appearance and fineness of the resulting product. The quality of ancient Egyptian earthenware was due in part to the availability of smooth Nile clays. The uniformity of the clay became especially important with the development of the potter's wheel in Mesopotamia during the period 4000–3000 BCE. The potter's wheel was a major artistic and technological achievement. It influenced form by allowing the creation of even circular shapes drawn up into tall cylindrical containers. The potter's wheel also enabled mass production of uniform and consistent ceramic wares.

The major earthenware products today are bricks, roof tiles, and ornamental terra cotta pots found at the local garden center and used for holding plants. Because earthenware is porous it is not the best container for carrying liquids; they leak. But the slow evaporation of water from the surface of the pot can actually keep the contents cool. So, earthenware makes very good containers for bottles of white wine. The same principle is at work in "desert fridges", which can be used to store food and medicines in regions with limited access to electricity such as rural parts of Africa. A desert fridge consists of two terra cotta pots of different sizes, placed one inside the other separated by a layer of sand.

The second general class of pottery is stoneware. Stoneware has similar ingredients to earthenware but is heated to a much higher temperature (1300–1400 °C). During this process the clay actually melts and on cooling produces a glass, which fills in all the pores. So, stoneware is not only able to hold liquids without leaking, its much stronger than porous earthenware.

Traditional stoneware pots were generally gray, but the color can vary from black to red, brown, and gray. The most desirable stoneware was white, because it represented a level of perfection in terms of purity of materials and consistency of processing that was difficult to achieve.

Fine white stoneware was first made in China as early as 1400 BCE (this is during the Shang Dynasty; 1600 BCE–1046 BCE). The sophisticated kilns designed by the Chinese provided excellent heat retention, an even heat distribution, and allowed the clay to be fired at higher temperatures than those attainable in other countries.

Johann Friedrich Böttger and Count Ehrenfried Walther von Tschirn-haus produced the first European stoneware, which was a brownish red, in Germany in 1706: only a mere 3,000 years after the Chinese! The Count was an expert in working at high temperatures and had developed a system, using glass lenses and mirrors, that concentrated and focused sunlight. He held the European record during the early eighteenth century for the highest temperature that had been obtained in a laboratory: an unprecedented 1436 °C, which was enough to melt a mixture of lime and sand.[4]

In 1767, sixty-one years after the success of Böttger and von Tschirnhaus, Josiah Wedgwood produced black stoneware that he called basalte and white stoneware colored by metal oxides that was called jasper or jasperware. In nature, jasper is a semiprecious gemstone that occurs in a range of colors, the most common being red, yellow, brown, and a pleasant pale green. One form of Wedgwood jasperware that is very popular is a turquoise green, which resembles the color of the natural stone. The attractive and highly recognizable blue jasperware is colored using cobalt oxide. (Cobalt oxide is also used to color all dark blue glass.) Wedgwood was so excited by his new blue ceramic body that he wrote to his partner: "The only difficulty I have is the mode of procuring and conveying incog (sic) the raw material.... I must have some before I proceed, and I dare not have it in the nearest way nor undisguised." Trade secrets have always been a critical part of the intellectual property of companies that manufacture ceramics.

Shortly after graduating from university I worked as a research scientist for Cookson Group, a multinational chemical company that owned a ceramic color manufacturer in Hanley (one of the UK's "Potteries".)[5] Our laboratory spent many months trying to reverse engineer ceramic colors obtained by various direct and indirect methods from our competitors (including items pulled from their garbage!). The one color that we could never duplicate even after many hundred attempts was the quite intense, yet delicate, manganese pink! Manganese, like cobalt, is a transition metal. These elements are the source of color in the majority of ceramic pigments.[6]

Porcelain, white, thin, and translucent, is a very special type of stoneware and was invented by the Chinese and produced during the T'ang Dynasty (618–907 CE). It is by far the finest and most highly prized form of traditional ceramic. If you tap a porcelain teacup with a spoon it will produce a metal-like ringing sound. Like stoneware, porcelain is fired at high temperatures, but stoneware is not usually white nor translucent because it does not use the same high purity ingredients. Porcelain is made from three ingredients: kaolin (a special clay known as China clay), feldspar (called petuntse by the Chinese, which forms a liquid at high temperature), and quartz in the form of sand. When the three ingredients are heated to 1300 °C and then cooled a complex structure forms where large particles of quartz are surrounded by a continuous sea of rigid glass.

Kaolin is the critical ingredient in porcelain. The name comes from the Chinese term "kauling" meaning high ridge and corresponds to the name of the mine in Kiangsi Province in eastern China, where the special clay was mined centuries ago. In addition to Chinese kaolin, there are major deposits in southwest England, around Cornwall, and in Ukraine. The major commercial source of kaolin in the United States was formed 50 million years ago and occurs as a continuous belt stretching along the ancient coastline from central Georgia northeast to the Savannah River area of South Carolina. The state of Georgia produces over 7 million tons of kaolin each year. Kaolin is not only used in the production of ceramics it is also found in paint and rubber. However, the largest use is not in ceramics it is in the production of paper. It is a key ingredient in making "glossy" paper glossy. The gloss and printability of the paper is determined by the size of the kaolin particles and how they are dispersed within the paper.

A few pieces of Chinese porcelain had trickled via the trade routes into Europe by the seventeenth century. The Portuguese had a trading center at Macao at the mouth of the Canton River. Following the establishment of the Dutch East India Company in 1609 a flood of oriental wares appeared in European markets. The European potters working with their crude earthenware were amazed by this very high-quality ceramic coming from Asia. Busily, they set about trying to reproduce what the Chinese had accomplished. They mixed different compositions and these attempts led to several grades of porcelain, but each lacked the translucency of true porcelain. It is not unexpected that one of the groups working on high temperature stoneware would unravel the secrets of porcelain. Eventually, in 1709, our friends Böttger and von Tschirnhaus figured out that the choice of clay was critical and for their European porcelain they used special white clay they brought from Colditz in Saxony. A factory was soon set up in Meissen, on the River Elbe about 70 km to the east of Colditz, for the production of stoneware and in 1710 it produced Europe's first porcelain. Unfortunately, the Count had died before the factory was even built and Böttger would die in 1719 aged only thirty-seven.

Bone china is an English version of porcelain produced by adding 50% animal bone ash to the basic porcelain recipe. Early versions of bone china were made by Thomas Frye at the Bow Porcelain Factory in London's east end, which was close to the cattle markets and slaughterhouses in the neighboring county of Essex. Firstly, the bones (the shin and knuckle from beef cattle apparently work best) must have all the tissue removed by boiling, which was an incredibly mucky and smelly process. The fatty residue was sent off to make glue. The boiled and stripped bones were crushed, burned, and turned into a fine ash.

The first successful commercial bone china formula was devised by Josiah Spode (son of the already well established potter Josiah Spode, senior) at the end of the eighteenth century. The structure of bone china is quite different from that of porcelain because it contains calcium phosphate, which comes from the animal bones and forms the mineral whitlockite (tricalcium phosphate) during firing. This phase improves the strength, translucency, and gives a creamy whiteness to the final product. Animal bone ash is still used in making bone china today. The English import a lot of their bone ash from Argentina, which is one of the largest cattle producers in the world.

Bone china was originally marketed as "English China" or "Spode China." The quality of "English China" was not lost on the British royal family and in 1806 the Prince of Wales appointed Spode 'Potter and English Porcelain Manufacturer to His Royal Highness' and asked him to supply the royal household with bone china and other ceramic products. Spode's factory was one of the main producers of blue-and-white printed pottery and popularized the "willow pattern," an example of which stood proudly for many years in my grandmother's—like many other English grandmother's—china cabinet. The main features of the pattern are the bridge with three people crossing it, the eponymous willow tree, the boat, the main teahouse, the two birds and the fence in the foreground of the garden.

Before we look at how the production of ceramics became a major global industry, it is relevant to take a small but important diversion and talk about glazes. This topic could have been in Chap. 6, the chapter on Glass because although the terms are not interchangeable, a glaze is a glass. Because the history of glazes is so intimately connected to the development of ceramics they are described here. Glazes added not just a new function but a whole new dimension to the aesthetics of pottery.

One of the very important properties of glass is that it behaves increasingly like a liquid the hotter it gets. This behavior is unlike that of a crystal, which is a solid up until it spontaneously transforms into a liquid at its melting temperature. At the temperatures used to fire ceramics a glaze is able to flow and in early applications of glazes they could be used to seal the pores in earthenware pots allowing them to retain water and become an indispensable material of life.

Glazing was introduced by the Egyptians around 3000 BCE and involved coating the fired pots with a solution consisting of finely ground sand mixed with sodium salts (sodium chloride, common table salt, worked well) or plant ash (which contains a large amount of sodium). Salt-tolerant plants of the genera *Salicornia* and *Salsola*, which are found in the deserts of Egypt and the Middle East, were widely used in the production of early glazes. After application of the glaze the pots needed to be reheated during which time the sand and salt particles fused together into a uniform protective glassy layer. In some cases the salt was thrown onto the pot while it was still in the kiln at high temperature. The flowing glaze produced interesting and unique decorative flow patterns.

The two most important types of glaze are the transparent lead glaze and the closely related opaque white tin glaze. The Chinese developed the world's first lead-containing glazes during the Warring States period (475–221 BCE). Lead compounds were widely available and among their many applications they were used in cosmetics, as pigments, and as a whitewash on the walls of important buildings such as government offices. Women used lead carbonate, a white powder also known as ceruse and white lead, as a cosmetic to hide facial flaws and create a pale uniformity over their skin. Litharge, also called yellow lead or massicot, was at times applied to the faces, particularly foreheads, of Chinese ladies to create popular "yellow brows."[7]

In a glaze lead has two important roles. Firstly, it decreases the softening temperature of the glaze, which allows the firing step to be at a lower temperature. Secondly, the presence of lead increases the refractive index of the glaze. The refractive index is a measure of the degree to which light is "bent" as it passes through a transparent material. Diamond, for example, has a very high refractive index, which is why the stones sparkle so brightly when professionally cut and polished. A high refractive index also means that more light is reflected from the glazed surface of a ceramic into the eye of the observer. In glazes, which are oxide glasses lead is present as the oxide (PbO). The lead oxide content of the early Chinese glazes was about 20%. During the Han dynasty (206 BCE–200 CE) higher lead oxide contents were typical, even up to as high as 60%. At these quantities, the glaze would have been easy to melt and have produced a product that was dazzlingly glossy.

A major problem with lead glazes is that acidic foods and liquids, such as lemons and wine, can leach out lead from a lead-glazed container. The body then ingests the lead. The cumulative effects of lead were certainly not appreciated by ancient Chinese and European civilizations and it was only relatively recently when organizations such as the U.S. Food and Drug Administration (FDA) developed standards for acceptable amounts of lead that could be "safely" leached from glazed cups and bowls. The standard test involves measuring how much lead is released when a glazed item is filled with 4%

acetic acid (essentially vinegar) at room temperature (fixed at 20 °C or 68 °F) for one day. The lead release from Eastern Han Dynasty (206 BCE–220 CE) lead glazes has been measured to be between 40–120 ppm (parts per million). The FDA limit for lead release from small tableware is only 2 ppm![8] As a word of caution, ceramics sold outside of the United States may not meet FDA standards—so buyer beware if you use them for anything other than ornaments!

Some historians, including the late Australian writer Jack Lindsay, believe that lead released from glazes helped poison a number of Roman nobility thus contributing (together with lead from water pipes) to the fall of the Roman Empire. [5] Early stages of lead poisoning include constipation and anemia, but at higher concentrations lead causes serious mental problems. At toxic levels, sterility is common and infant mortality rates can be very high. In an article "Lead Poisoning in the Ancient World," published in the journal *Medical History*, Dr. H.A. Waldron of the Department of Anatomy at the University of Birmingham in the UK described a possibly more significant source of ingested lead that came from lead-coated cooking pots that were widely used for preparing wine and grape syrup. [6] Dissolved lead helped to sweeten the wine.

Even during the nineteenth century, lead poisoning was a major cause of death for workers in the pottery industry. In the six small towns that made up the "Potteries" in England, 432 cases of lead poisoning were reported in 1897 alone. This was 45 years after a major British government report 'The Employment of Children in Factories,' which had described the 'harsh,' 'bleak,' and 'miserable' lives of young children working in the pottery industry and the prevalence of industrial disease. It only took another fifty or so years when in 1949 British regulations forbade the use of raw lead in glaze compositions. Consequently, British potters and those in many other countries switched to using "fritted" glazes.

A frit is granulated glass that has been produced by melting its constituents together and once a homogenous melt has been achieved pouring the hot molten glass very quickly into a trough of cold water. As the hot glass hits the cold liquid it spits and crackles. Amid the clouds of steam, a mass of small glass fragments is formed.[9] Lead-containing frits are widely used in the ceramics industry because they help bind the toxic component into an inert silica matrix reducing the chance that it will be leached out. As a result of fritting and appropriate health and safety regulations lead poisoning was eventually eliminated in the Potteries and most other industrial potteries.

In addition to the lead glaze, the other important glaze was the opaque tin glaze first produced in Iraq in the eighth century. The tin glaze shares a lot in common with lead glazes, except that rather than being transparent the presence of tin oxide makes it opaque. Tin oxide is present in concentrations between 5–10% in the form of sub-micron sized particles that give the glaze a white appearance. [7] In the ninth century the tin-opaque glaze spread into Europe via the popular Spanish holiday island of Majorca after which it was later named *majolica*. Centers of majolica manufacture developed in Faenza in Italy (the Italians named their glaze *faience*) and in 1584 at the famous production center at Delft in the Netherlands (named appropriately *delftware*). During the sixteenth and seventeenth centuries, majolica was the principal luxury ceramic throughout Europe. Good quality examples of majolica, faience, and delftware can be very valuable and command high prices; just watch an episode of *Antiques Roadshow*!

Tin glazing became of industrial importance at the end of the nineteenth century with the growth of the ceramic sanitaryware industry. One of the most important sanitaryware products of the time was the toilet, which represented an important advance in public health. The first single piece ceramic flush toilet was manufactured by Thomas Twyford in 1885. This was at a time of "the great sanitary awakening" where human waste and filth was identified as a major cause of disease and a vehicle of transmission. [8] In London cases of smallpox, cholera, typhoid, and tuberculosis were at all-time highs. In New York smallpox and typhus were rampant. Proper sanitation facilities were critical because they promoted health through appropriate disposal of human waste.

In the twentieth century some of the great artists such as Pablo Picasso and Marc Chagall experimented with tin glazed pottery as an artistic medium. Examples of their earliest tin-glazed pieces can be seen in the Museo Nazionale delle Ceramiche in Florence.

Glazes can be colored by adding to the formulation specific compounds such as cobalt oxide, which as noted earlier produces a deep blue color. In celadon glazes that were first produced 3,500 years ago the color can vary from light blue all the way to yellow green depending upon the amount of iron oxide (Fe_2O_3) added to the glaze.

A recent find on display at the Oriental Metropolitan Museum in Nanjing, Jiangsu Province is a celadon jar made in the third century with under-glaze painting featuring winged figures of Taoism. The significance of this jar, shown in Fig. 3.4, is that it represents the earliest porcelain ware painted with patterns *under* the glaze, rather than applied *on* the surface. It challenged the previous notion that the technology of under-glazed porcelain started in the seventh century. A more recent example of celadon pottery, this time from Korea, is the small water container, 23 cm tall, given to former American President Harry Truman in 1946 by the government of the Republic of

Fig. 3.4 Underglazed red vase (*meiping*) with design of pine, bamboo and prunus from the Hongwu reign (1368-1398) with a lid. This vase, which is on display in the Nanjing Museum is the only one of its kind intact in the world

Korea. It is now valued at $3 million. Additions of up to 8% of iron oxide to the glaze formulation creates a thick, lustrous dark brown almost black color. This is the Tianmu (in Japanese *temmoku*) glaze developed during the Sung Dynasty (960–1279 CE). Tianmu was popular for tea bowls, where the dark brown on the wall of the bowl would break to a lighter brown on the rim.

Large-scale manufacturing of ceramics became an important industry in Europe during the eighteenth century. It transitioned from what was largely a craft institution where small groups of potters used readily available sources of clay and other raw materials to supply local markets to an organized industry employing thousands of people using some of the most advanced manufacturing methods. The major sites of this transition were in Europe: at Sèvres in France, Meissen in Germany, and most significantly, the Potteries in Staffordshire, England.

Claude-Humbert Gérin and brothers Robert and Gilles Dubois established the national porcelain factory of France at Vincennes in 1740. Madame de

Pompadour, the mistress of Louis XV, had a keen interest in porcelain and was able to help obtain royal patronage for the factory. It moved to Sèvres, located between Paris and Versailles, in 1756 and became the most famous porcelain house in Europe. The factory focus was on the manufacture of high-quality porcelain for the nobility. One of the most famous commissions was an 800-piece dinner service (the "cameo service") for Catherine II (Catherine the Great) of Russia that took three years to complete. Some of the most elaborate porcelain to come from Sèvres cost as much as silver or gold and the French royal family often gave Sèvres porcelain as gifts to visiting foreign dignitaries. During the French Revolution, the Sèvres factory was taken over by the state in 1793.

The Meissen factory was established near Dresden in Germany in 1710, under the direction of Böttger. Initially the factory produced products solely for the royal family. It moved to producing more commercial wares in 1713. The increasing popularity of tea, coffee, and chocolate among the wealthy created an added demand for high quality porcelain containers to hold these luxury beverages.

The Potteries in Staffordshire, England consisted of six towns: Tunstall, Burslem, Hanley, Stoke-on-Trent, Fenton and Longton. There had been many small potteries in this area since the late seventeenth century. By the mid-eighteenth century these potteries were employing a large number of workers. For example, in Burslem there were nearly 150 separate potteries making various kinds of stoneware and earthenware, which together employed nearly 7,000 people. From humble beginnings, Staffordshire grew to become one of the largest and probably the most influential ceramic manufacturing center in the world with 50,000 men, women, and children making cups, plates, and saucers for dinner tables around the world. English writer Arnold Bennett, whose ashes lie in Burslem cemetery, was a great chronicler of life and times in the Potteries. In *The Death of Simon Fuge*, a wonderfully amusing short story, Stoke-on-Trent (called Knype by Bennett) is described as having "a squalid ugliness on a scale so vast and overpowering that it became sublime". [9] But there is the realization from the smug and effete narrator, a visitor to the Potteries from the British Museum in London, that it is easy to forget that "coal cannot walk up unaided out of the mine, and that the basin which he washes his beautiful purity can only be manufactured amid conditions highly repellent."

Tea was an important factor in the growth of the English ceramic industry during the eighteenth and nineteenth centuries. In the 1840s, formal tea parties became popular. A quick afternoon 'cuppa,' which had often been supped standing up, became a full-blown production. These parties required cups and saucers, plates for biscuits and cakes, muffin dishes, a slop bowl into which waste tea leaves were poured, a ceramic stand to protect the polished surface of a wooden table from the heat of the teapot, and many other essential bits and bobs. All of which had, of course, to be matching. The demand from the Victorian public became enormous and the Staffordshire potters were happy to oblige in meeting it.

Staffordshire is a long way from England's major metropolitan areas such as London, Bristol, Manchester and Birmingham. Its development as the prominent pottery center in England was in large part due to the availability of coal as a fuel for the kilns. (The King's Navy placed an overriding demand on the nation's timber supplies to build more and more war ships.) Coal was abundant in this area as was, and still is, a good source of clay. At the beginning of the nineteenth century it took 2.5 lb (about 1 kg) of coal to fire one pound (less than 0.5 kg) of clay. So the process of producing ceramics required large amounts of energy and was inefficient. Even with the local abundance of fuel, efficiency was a concern and has remained a major issue for the ceramics industry. The proximity of raw materials, particularly coal, provided an economic advantage for the Potteries over other possible rural locations in England that were still relying on the diminishing supply of timber.

In North America the origin of pottery production occurred in regions where there were abundant supplies of earthenware clay deposits and the wood needed as fuel for the kilns. The abundance of these raw materials underfoot and in the forests of Virginia was a factor in the English settling in Jamestown in 1607. There is evidence that pottery production began there around 1625. Various ethnic groups played important roles in the new settlements. For example, in the memoirs of Captain John Smith he describes how many of the settlers from Poland who arrived in Jamestown on October 1, 1608 were hired to make glass. [10] The immigrant potters produced a variety of articles for use in the home and the farm: storage jars, pitchers, bowls, mugs, milk pans, porringers. These were all made on the wheel using the local earthenware clay.

Clay was the plastic of the ancient world. It's versatility, durability, and ability to be shaped made it an essential material in the development of civilization. Our ancestors, like potters today, learned how to take a shapeless mass and turn it into some of the most beautiful and striking objects ever created. Whereas flint was always shaped by subtraction, removing flakes until the final shape was obtained, clay could be shaped by addition, building layers of raw material creating forms that are increasingly complex and wonderful.

Clay was the first material whose use necessitated a change of state, from a fluid plastic mass to a hard ceramic, which ultimately led to the development of high temperature furnaces that were critical for working with metals.

Clay has remained a material that is as important now as it was in the ancient world. Over 40 million tons of kaolin, a critical ingredient that allowed the Chinese to discover the secret to making porcelain, is mined each year. Much of this still goes into the production of fine ceramics. Many of the defining applications for clay such as containers and bricks have remained. But over time new applications have emerged. Today about 10% of all the clay produced is used as a catalyst or catalyst support to enhance chemical reactions such as the production of synthesis gas (a mixture of hydrogen and carbon monoxide) from methane, the primary constituent of natural gas. Today 95% of all the hydrogen produced in the United States is made from natural gas. This resource will be critical in any transition towards a hydrogen economy.

Notes

1. In the Universe the lightest elements hydrogen and helium are the most abundant. These elements were formed during the Big Bang. The remaining naturally occurring elements from carbon onwards were formed in the fiery depths of stars and released in violent explosions. Heavier elements such as titanium required the conditions created when massive stars, 1,000 times bigger than our sun met their explosive deaths. Even heavier metals such as gold were born in the powerful collisions between neutron stars.
2. There are relationships between processing, structure, and properties of materials. Processing affects structure; structure determines properties; we can design for certain properties by processing to obtain a specific structure.
3. The color of a hot object is a good indication of its temperature. Barely visible red corresponds to 525 °C; a bright red, turning orange would indicate a temperature of 1,000 °C. When bright white the object has reached 1400 °C.
4. A modern version of von Tschirnhaus's furnace is known as an arc-image furnace. It operates in the same way by focusing light using parabolic mirrors. Using an arc-image furnace it is possible to achieve temperatures > 2,500 °C, which is enough to melt most materials.
5. The original Cookson Group was formed in 1704 by Isaac Cookson. In 1924 it merged with Locke, Lancaster, and W.W. & R. Johnson & Sons to become part of the Associated Lead Manufacturers, Ltd. Associated Lead Manufacturers, Ltd made a series of acquisitions to expand the market for their lead products, mainly the colorful oxides that were used as paint additives. A 1961 advertisement for Associated Lead Manufacturers, Ltd., boasted: "Wonderful what LEAD

can lead to." (Their capitals.) In 1967 the company, after further acquisitions and amalgamations, was renamed Lead Industries Group primarily to reflect that the range of products it offered was broader than the manufacture and sale of lead compounds. It included other metals such as antimony and ceramic products such as titanium dioxide and zircon. The latter is used to make furnace bricks for the glass and iron and steel industries. By 1969 only a fifth of the company's profits came from lead paint. Shortly after I joined Lead Industries Group it was decided that the troublesome "Lead" should go and in 1982 the company was renamed the more innocuous Cookson Group, plc.

6. Cobalt oxide produces an intense blue when added to glazes. The addition of a small amount of chromium oxide gives the glaze a green color. Many of the early pigments produced colors that were unstable at the high temperatures used for ceramic processing. By trapping the color inside a transparent shell, it was possible to make more stable stains. For example, the zircon-iron pink is produced when tiny crystals of hematite (an oxide of iron with a blood red color) are encapsulated in zircon (an oxide of silicon and zirconium). By replacing the iron with the rare-earth element praseodymium, a bright yellow color was formed. Trapping a mixture of cadmium, selenium, and sulfur inside a zircon shell produced a brilliant and intense red.

7. Schafer, Edward H. (1956) The early history of lead pigments and cosmetics in China, *T'oung Pao*, Second Series, 44 (Livr. 4/5), 413–438. Lead compounds were made in China possibly as early as the Shang Dynasty, but definitely during the Han Dynasty. In addition to white lead (lead carbonate, the compound $PbCO_3$) and yellow lead (litharge, the compound PbO) there are other colored lead compounds. One of the most important is red lead (minium, the compound Pb_3O_4). The preparation of red lead directly from metallic lead was understood during the Zhou Dynasty. The process has not changed significantly over the past 2,000 plus years. I briefly worked for Associated Lead's manufacturing division in Blyth near Newcastle-upon-Tyne making red lead from metallic lead using a rotary furnace.

8. CPG Sec. 545.450 Pottery (Ceramics); Import and Domestic—Lead Contamination. FDA report https://www.fda.gov/ucm/groups/fdagov-public/@fdagov-afda-ice/documents/webcontent/ucm074516.pdf Accessed 1 February 2019.

9. Between 1981 and 1984 while I was at Cookson Group Central Research Laboratories in Perivale, UK I spent a lot of time developing potential "low sol" lead silicate frits for glazes.

References

1. Vandewettering, Kaylea R. (2015). Upper Paleolithic Venus figurines and interpretations of prehistoric gender representations. *PURE Insights, 4*, Article 7.
2. McDermott, L. (1996). Self-representation in upper Paleolithic female figurines. *Current Anthropology, 37*, 227–275.
3. Cooper, E. (2000). *Ten Thousand Years of Pottery* (p. 10). London: The British Museum Press.
4. Conway, G. (1976). Pottery reduction firing with a fuel-burning kiln. *Leonardo, 9*, 89–93.
5. Lindsay, Jack (1968). *The Ancient World: Manners and Morals*. New York: Putnam.
6. Waldron, H. A. (1973). Lead poisoning in the ancient world. *Medical History, 17*, 391–399.
7. Tite, M. S., Freestone, I., Mason, R., Molera, J., Vendrall-Saz, M., & Wood, N. (1998). Lead glazes in antiquity—methods of production and reasons for use. *Archaeometry, 40*, 241–260. The tin oxide (SnO_2) particles in the glass are about 500nm in diameter and make the glaze appear white.
8. Winslow, C.-E. (1923). *The Evolution and Significance of the Modern Public Health Campaign*. New Haven: Yale University Press.
9. Bennett, Arnold (1912). *The Matador of the Five Towns and Other Stories*. New York: George H. Doren Company. The Death of Simon Fuge is the fifth in this collection of Bennett's Potteries stories.
10. *The Complete Works of Captain John Smith (1580–1631)* in Three Volumes, Philip L. Barbour (ed) http://www.virtualjamestown.org/exist/cocoon/jamestown/fha-js/SmiWorks1 Accessed February 2, 2019

4

Iron—*The Material of Industry*

The Severn is the longest river in the United Kingdom. Its 220-mile journey to the sea begins in the Cambrian Mountains of mid Wales. Spanning the notoriously volatile river and the deep Ironbridge Gorge in Shropshire is a landmark that represents the epicenter of the Industrial Revolution. Located in a part of the English countryside rich with easily accessible raw materials—coal, iron ore, clay, and limestone—needed to power a new type of industry the Iron Bridge was both a bold technical achievement and represented the aesthetic possibilities of cast iron. Built in 1779 and opened New Year's Day 1781, the Iron Bridge, shown in Fig. 4.1 made using nearly 400 tons of cast iron was the perfect symbol of the Enlightenment, when technological innovation promised limitless possibilities for the improvement of life [1].

It was the innovations that came from the foundries around Ironbridge that revolutionized the iron industry and laid the foundation of modern iron and steel production. Iron is by far the most widely used of all metals. Annual production of iron and steel is over 3,000 million tons. By comparison world aluminum production is about 65 million tons.[1]

This chapter describes the properties and applications of three forms of iron: cast iron, wrought iron, and the cheapest and most extensively used iron alloy: steel. How did iron, and then steel, reach its level of technological and societal importance? Without steel we would not have had the growth of railways, which were critical to the expansion of the American west. City skylines across the world from Shanghai to Seattle would look flat

© The Author(s), under exclusive license to Springer Nature Switzerland AG 2021
M. G. Norton, *Ten Materials That Shaped Our World*,
https://doi.org/10.1007/978-3-030-75213-2_4

Fig. 4.1 The Iron Bridge across the River Severn is one of the landmarks of the Industrial Revolution. The bridge is an engineering achievement and demonstrated the aesthetic possibilities of cast iron. It was described by Viscount Torrington in 1784 as 'one of the wonders of the world.' (Image courtesy of Historic England, reproduced by permission.)

and uninteresting without the skyscrapers that rely on steel beams for their construction. Most international shipping is done using steel-hulled ships. How would commodities, such as oil, be transported in large quantities all over the world without oil tankers and container ships? This chapter explains how iron became the most indispensable metal of modern civilization.

Like several other materials in this book—gold, aluminum, and silicon—iron is an element, a primary constituent of matter. Each element in the Periodic Table of Elements is distinguished by its atomic number—the number of protons in the nucleus of its atoms. Iron has atomic number 26 and is one of the so-called "transition elements," which make up the largest block of the Periodic Table and include vanadium, chromium, and manganese.[2] These elements, and many others, can be added to iron to form a range of alloys, each with its own unique properties. Chromium, for example, is what makes stainless steel, "stainless." Vanadium steel is used in cars for axles, crankshafts, and gears because it provides superior strength. Alloys of iron and nickel were used, albeit to a very limited extent, in China,

Iraq, and Egypt even before the Bronze Age. In China the earliest form of iron dates back to the Shang Dynasty that ruled from 1600 through 1100 BCE. These alloys were naturally found in meteorites that had long ago collided with the Earth. Meteoric iron contains about 10% of nickel along with small amounts of cobalt and manganese.

Some very large examples of iron meteorites can be seen, and touched, at the American Museum of Natural History in New York City. One of the largest of all is the Willamette meteorite that weighs in at an impressive fifteen tons (over 13,600 kg.) Ellis Hughes, a Welsh immigrant, came across the enormous meteorite in 1902 on land that belonged to the Oregon Iron and Steel Company (located in Oswego, not far from the banks of the Willamette River). Hughes carefully, over several months, dug up the meteorite, loaded it on to a cart and moved it onto his property. The enterprising Hughes then charged visitors a quarter to view the meteorite. The Oregon Iron and Steel Company became suspicious about where the meteorite had been found and sued Hughes. The company prevailed in the legal case and eventually the meteorite was sold. The new owners gave it to the American Museum of Natural History where it remains to this day.

The Hittite king Murshelish III had implements forged from meteoric iron as far back as 1400 BCE. His empire, which is now Syria, was on an old trade route from Iran and Iraq to Egypt, where some of these iron objects eventually found their way. The Egyptian pharaohs were also very interested in iron. When Tutankhamen's tomb was opened in 1925 by archaeologist Howard Carter, in addition to the usual riches of gold and jewels, there were a number of items made of iron: a dagger, a blade, a bracelet, and a miniature headrest. The dagger has an iron blade with a gold handle, rock crystal pommel and an intricately decorated sheath. The iron blade has fascinated researchers since its discovery. Not only was ironwork rare in ancient Egypt because of its meteoric origin, but the metal had not rusted even though it is more than 3,300 years old.

The Egyptians prized iron in part because it was so useful—for example, it is much stronger and tougher than gold[3]—and in part because it was magical: it came from heaven. The earliest Egyptian word for iron is "benipe," which is translated as "metal from the sky" or "star metal." Recent analysis of the composition of the iron blade, shown in Fig. 4.2, found in Tutankhamen's tomb showed that it contains a large amount of the element nickel and a smaller amount of cobalt. This combination provides strong evidence for its extraterrestrial origin. By analyzing known meteorites within a 2,000 km radius of the Red Sea coast in Egypt, researchers believe that they have identified the exact meteorite that the iron came from. That meteorite, named

Fig. 4.2 The iron dagger of King Tutankhamun. Color picture of the iron dagger (Carter no. 256K, JE 61585) with its gold sheath. The full length of the dagger is 34.2 cm. (Source: Daniela Comelli, Massimo D'orazio, Luigi Folco, Mahmud El-Halwagy, Tommaso Frizzi, Roberto Alberti, Valentina Capogrosso, Abdelrazek Elnaggar, Hala Hassan, Austin Nevin, Franco Porcelli, Mohamed G. Rashed, and Gianluca Valentini, (2016) "The Meteoritic Origin of Tutankhamun's Iron Dagger Blade," *Meteoritics & Planetary Science* 51 (2016): 1301-1309. Reproduced with permission Wiley.)

Kharga, was found at the seaport city of Mersa Matruh, 150 miles west of Alexandria [2].

Meteoric iron is not abundant enough to have provided a useful nor widespread source of iron and certainly would never have been sufficient to usher in the Iron Age. But there is an abundance of iron in the Earth's crust. Iron is also the principal metal in the core (the inner core is mostly solid iron and the outer core contains liquid iron and sulfur). During the formation of our planet some 4.5 billion years ago iron-containing dust condensed into a ball. As this fiery ball cooled, the heavy elements like iron sank towards the center where they solidified forming the inner core. Iron is also the second most abundant metal in the Earth's crust (5% by weight behind aluminum's 8%). Iron in the outer layer of the planet reacted with oxygen to form a number of different iron oxides. Like most other metals, iron exists predominantly as a constituent of minerals, where it is tied up with oxygen as an oxide or with sulfur as a sulfide. The oxides are our mineral sources of iron. Minerals that are used commercially as sources of metals are called "ores." The common ores of iron are all oxides, and they are among the most abundant substances on Earth.

The three important oxides of iron are magnetite, hematite, and limonite. Magnetite is also known as lodestone, from the old English word *lode*, which meant *guide*. Magnetite was the first magnetic material to be discovered. In its natural state magnetite is permanently magnetized and its power was well known in ancient times. Socrates is said to have played with pieces of lodestone and found that they magnetized iron rings: much like a magnet-tipped screwdriver can magnetize steel screws. Chinese compasses made during the Han Dynasty (206 BCE—220 CE) used a ladle carved from lodestone resting on a polished bronze plate where it was free to rotate under the influence of the Earth's magnetic field.

While magnetite is black, hematite is usually a reddish-brown reminiscent of dried blood. The word hematite is based on the Greek *haima* for blood. The red color of blood comes from iron atoms that are in the same oxidation state as they are in hematite: in both cases the iron has the same number of electrons. Limonite is a less pure hydrated form of iron oxide and usually has a dull red or brown appearance. It is the common coloring agent found in soil and on the surface of Mars.

The most important source of iron ore in North America is a mixture of magnetite and some limonite in what is called the Mesabi Iron Range, which extends for about 120 miles in St. Louis and Itasca Counties, Minnesota, and is in places up to three miles wide. During the 1920s to 1950s this was the most productive iron field in the world. The iron ore was a critical raw material for making weapons for World War I and World War II. At its peak in 1953, 76 million tons of ore were produced in just a single year. The iron produced from this range was a key factor in making America a post-war industrial giant. But, by around 1955 many of the mines located in the Mesabi Iron Range had become depleted of high-quality ore and production decreased.

Iron ore contains lots of oxygen as well as impurities, which lower the metal content. Iron ore is converted into metallic iron by a process called smelting, which involves removing the oxygen using some form of carbon such as charcoal or coal. The carbon reacts with air to form carbon monoxide, which bonds with another oxygen atom from the iron ore to form carbon dioxide (CO_2).

The origin of smelting iron ore, like many early technologies, remains uncertain. One very likely possibility is that it may have been smelted first in Africa where primitive methods can still be found in isolated communities. By 1500 BCE the art of smelting iron ore was practiced by people living on the slopes of the Caucasus Mountains facing the Black Sea. From there the technology spread to nearby Iran and Iraq. The extraction of iron from its ores was subsequently independently discovered—forgotten and rediscovered—throughout Europe and Asia.

The earliest smelting furnaces did not produce molten iron. Iron has a melting temperature higher than gold, silver, and copper, and because of their simple design the ancient furnaces simply could not get hot enough to melt it. As a result, early metal workers were not able to produce a refined or highly pure form of iron. What they did produce was a spongy impure mass called *bloom*. Bloom could be beaten into shape and used to create useful objects such as small spearheads for hunting, which were sufficient to demonstrate the potential of iron. Thus, dawned the Iron Age. Only a relatively short interval (less than 2,000 years) separated the Bronze Age from the widespread use of iron. So, after 2 million years of using naturally occurring materials such as seashells, bone, and stone, in just a few thousand years our ancestors had begun to smelt ores and use metals.

As is the case with the discovery of many new materials, one of the earliest applications for iron was to make weapons [3]. This is seen in the number of daggers, swords, axes, blades, and arrowheads made of iron that have been discovered in addition to iron tools. The rise of the Assyrians, Babylonians, and Persians has been attributed, in part, to their superior weaponry. The Assyrian Empire, which lasted from 1100 to 600 BCE made use of iron blades attached to bone hilts to make powerful swords. The Persian smiths were making bronze-hilted swords around the same time. Although pure iron is not significantly harder than bronze, it is possible that some of the iron produced during this time did contain sufficient carbon and the necessary structure induced by the right heating and cooling cycles to make an iron that was unequivocally harder than any bronze. Certainly, sources of iron are far more abundant than copper and tin, the two primary constituents of bronze, and did not rely on elaborate trade routes. As a result iron weapons would be more prevalent than bronze, giving a numerical advantage, and they were less expensive.

By about 1000 BCE, furnaces capable of melting mixtures of iron and carbon were developed. The iron-carbon mixture melts at a lower temperature than pure iron. The Chinese were the pioneers of this technology, which spread to India before the Christian era and was independently discovered in Europe during the fourteenth century (a delay of only 23 centuries!). The Chinese metal workers didn't necessarily have furnaces that could go higher in temperature than those in western Asia and Europe, but they were working with iron that had a very high carbon content after smelting. Pure iron melts at 1573 °C. If only 4% of carbon is added to pure iron the melting temperature drops by an amazing 421 °C or over 25%. A temperature of 1152 °C was certainly attainable by furnaces available during the early part of the Iron Age. So, the Chinese made iron with a high carbon content, which could be melted and poured into molds to make things. Consequently, this form of iron is called *cast* iron. Because of their technological lead and genius, the Chinese became the world's experts in casting iron. The impressive 40-ton Cangzhou lion in Hebei province, which dates from 930 CE is the largest iron casting artwork in the world.

Cast iron made possible many technological innovations that shaped our world. In 1817 English inventor Richard Roberts designed a heavy-duty lathe made entirely of cast iron. Prior to this the processes and tools used for shaping of wood and metal had changed little since the Middle Ages. The Roberts' lathe could turn out metal parts faster and with a higher degree of precision than was possible using the existing methods in the early part of the nineteenth century. These standardized components made mass production possible and enabled further world shaping innovations including steam engines, locomotives, and power looms. Steam engines powered mills and factories, freeing them from the need for a nearby water supply. Stephenson's "Rocket", which in 1829 pulled a carriage of passengers over a distance of 35 miles at an average speed of 12.5 mph started a revolution in transportation that has impacted almost every corner of the world. The power loom transformed the manufacture of cloth from a skilled craft into a mechanized industrial process.

Despite these undeniable engineering accomplishments, however, a major problem with cast iron is that it is very brittle. This limitation meant that it was not useful for making certain objects—including swords. During the period 400–200 BCE, approximately the same time when Syrian and Roman cultures were beginning to use iron for agriculture and items like iron pipes for glass blowing, the Chinese developed a method to make cast iron malleable enough to become useful to make weapons. They transformed their cast iron by a process called decarburizing, which removes some of the carbon. Carbon has an affinity for oxygen: the carbon–oxygen bond is one of the strongest in nature. So, heating up cast iron in air will cause some of the carbon to be removed from the iron as carbon dioxide (CO_2). If the decarburization process is controlled carefully then it is possible to make a sword that has tough edges (where more carbon has been removed) with a strong core (where less carbon has been removed).

Cast iron is at one end of a broad spectrum of iron compositions. At the other end of the spectrum is *wrought* iron, which is essentially pure iron (with no carbon, nor much of anything else, added). Wrought iron is softer than cast iron and can be shaped by beating with a hammer. The bloom that was produced in the early iron smelting furnaces and subsequently beaten into shape is a form of wrought (or "worked") iron. The constant and repeated hammering of the bloom would force out unreacted pieces of ore and charcoal leaving behind a fairly pure product.

Wrought iron is very tough and won't break easily. However, it is more ductile than cast iron because it does not contain carbon. This behavior limits the range of applications for wrought iron. So, the applications for both cast iron and wrought iron are limited because of their properties. In one case it is too brittle, in the other case it is too soft.

Despite the acknowledged limitations of wrought iron, it could be, and was, used as a very effective construction material as illustrated by two very impressive examples: the S.S. *Great Britain* and the Eiffel Tower. The *Great Britain* was the first oceangoing propeller-driven ship and at almost 3,000 tons was the heaviest ship that had ever been built. It was also the world's first luxury liner. The famous British engineer Isambard Kingdom Brunel built the ship over the four-year period from 1839 to 1843. The hull was constructed of wrought iron plates (measuring 2 m × 0.65 m) that were riveted to metal frames. For twenty-four years the *Great Britain* carried passengers between Liverpool and Melbourne, Australia: from one end of the Earth to the other. The majestic ship now works as a museum from its mooring at the Great Western Dockyard in Bristol.

The Eiffel Tower in Paris was erected in 1889 as a "temporary" exhibit for the World's Fair. It is made from 7,300 tons of wrought iron held together with 2½ million rivets and stands 984 feet tall. Despite its temporary beginnings over 250 million people have visited the tower and with 7 million visitors a year it is the most visited paid-for monument in the world. At the time of its construction, it was the tallest human-made structure in the world. The Eiffel Tower beat out the Washington Monument in Washington D.C. by over 400 feet. After 41 years it lost the world title to the Chrysler Building in New York City, which is 1,046 feet tall.

Gustave Eiffel chose wrought iron over the other two choices, cast iron and steel. He felt that steel was too new and therefore more expensive than iron. Wrought iron, unlike brittle cast iron, can be rolled into plates or beams that can be easily assembled with rivets to form light and strong structures.

In this chapter we have used a number of terms to describe the mechanical properties of iron. Although we often use "strong," "hard," and "tough" interchangeably they actually have very specific meanings in engineering. Strong is the ability to resist bending; hard is the ability to resist being indented (such as by a nail); tough is the ability to resist breaking. Consequently, a material can be strong but not necessarily tough. A ceramic cup would be a good illustration of this property. It is difficult to bend the cup, but if you drop it on a hard floor it breaks very easily. On the other hand, many plastics are tough, but not particularly strong or hard. A good illustration would be a rubber band. It is easy to bend it, but you have to stretch it a very long way before it breaks. What we often are seeking in a material is one that combines strength and toughness.

For those groups making tough wrought iron there are ways to make it stronger. The first approach is termed work hardening. The basic idea is that all metals contain defects called dislocations—essentially, rows of displaced atoms. Their presence was first suggested in 1892 as "modifications in the structure" that be "transferred from one portion of the aether to another" [4]. The aether is basically a concept of medieval science used to describe everything that can't be described precisely. So that could include the structure of a crystal, the composition of the atmosphere, or the nature of space. When a metal bends—just think of bending a wire clothes hanger or a paper-clip—dislocations in the structure move to create a permanent deformation. As a force is applied to the metal, dislocations move from one location to another. When the force is subsequently removed, it is impossible for the dislocations to move back to their original position. Consequently, the metal has been permanently deformed. If we try to bend the clothes hanger or paper clip back, we can never return it to exactly the same shape: there is always a remaining kink.

If an excess of dislocations is introduced into the metal structure—by, for example, repeatedly hitting it with a hammer—it actually makes the material stronger, because there are now so many dislocations that they get in each other's way when they try to move, which causes a dislocation "traffic jam," slowing them all down. As a result, it now becomes more difficult for the metal to deform and so—by our earlier definition—it is now stronger. This is the principle behind work hardening.

Besides work hardening there are two other ways to strengthen iron: we can add hard particles to form a composite in a process called precipitation hardening (somewhat similar to adding aggregate to cement to form concrete), or we can cool the formed object very quickly. The easiest way to form hard particles within pure iron is to deliberately add a small amount of carbon—essentially doing the opposite to decarburization—which reacts with the metal to form cementite. When a moving dislocation hits a hard cementite particle it is stopped in its tracks, making the metal more difficult to bend. Particles of cementite are always present in cast iron and are what give it its brittleness. Cementite particles can also be formed, in smaller numbers, to strengthen wrought iron. There is evidence from examining the microstructure of ancient iron objects and finding cementite particles that carburization was used as early as 1400 bce.

Adding carbon to pure iron to strengthen it is still used today and is called "case hardening." Carbon monoxide is diffused into the iron producing a hard surface layer that is good for applications such as wear-resistant gears. The process was further developed and became eventually the cementation process, which dominated steel making in England in the eighteenth and nineteenth centuries. In the cementation process, wrought iron was packed into a chamber with charcoal where it was ignited and heated for several days. Small rods were periodically removed—like small tasters from a big block of cheese—to test for strength and when the desired performance had been achieved the process was stopped. It is worth pointing out that during cementation the iron was always solid—it was never heated to a high enough temperature to melt it.

The process of strengthening iron by cooling very rapidly, or quenching, began to be used in western Asia and Europe sometime around 1200–1000 BCE. Rapid cooling would have been counterintuitive to conventional wisdom, which was based on previous experience with bronze: bronze gets softer the faster you cool it. However, iron behaves in the opposite way. The rapid cooling of iron creates a new crystalline phase called martensite, which is very hard.

Martensite only forms under very rapid quenching: you need to go from 700 °C to room temperature in about 1 s. This rapid cooling was achieved by plunging the hot iron into a cool liquid. Water is an ideal choice, but reports indicate blood and urine also worked well as described in the book *Von Stahel und Eysen* (*On Steel and Iron*), published in Nuremberg, Germany in 1532: "Take the stems and leaves of vervain, crush them, and press the juice through a cloth. Pour the juice into a glass vessel and lay it aside. When you wish to harden a piece of iron, add an equal amount of a man's urine and some of the juice obtained from the little worms known as cockchafer grubs. Do not let the iron become too hot but only moderately so; thrust it into the mixture as far as it is to be hardened. Let the heat dissipate by itself until the iron shows gold-colored flecks, then cool it completely in the aforesaid water. If it becomes very blue, it is still too soft" [5]. Alternatively, "Take varnish, dragon's blood, horn scrapings, half as much salt, juice made from earthworms, radish juice, tallow, and vervain and quench therein. It is also very advantageous in hardening if a piece that is to be hardened is first thoroughly cleaned and well polished."

Exploring the range of properties possible with iron—tough, strong, brittle, hard, soft, malleable, all possible by controlling the chemistry and how the metal was processed, led to the development of what would become the most common and important engineering alloy of all time: *steel*. If cast iron was too brittle and wrought iron too soft, then steel could be made just right. Like cast iron, steel is an alloy of iron and carbon, but typically contains less than 2% carbon; half the amount in cast iron, and often contains other elements to create specific desirable properties.

The Hittites were responsible for the earliest form of steel, sometimes called "good iron" in old records, which was made around 1400 BCE. A hundred years after Tutankhamen, the Egyptian ruler Rameses II received an iron dagger blade from the reigning Hittite king, Hattushish III, together with a message complaining about his iron: "As to the good iron in my sealed house in Kissuwadna, it is a bad time to make iron, but I have written ordering them to make good iron. So far they have not finished it. When they finish it, I will send it to thee. Behold, now I am sending thee an iron dagger blade." [6]. The process to make good iron involved repeated heating and hammering of bloom. During this process some carbon—probably from carbon monoxide—found its way into the iron and strengthened it. As you can imagine, the balance in getting the right amount of carbon into the iron was critical. Too much and you had brittle cast iron, too little and you formed soft wrought iron, but just the right amount produced steel.

An interesting early method for making steel was the "wootz" process developed in India. This process is very similar to cementation and involved heating iron and charcoal in a sealed crucible. Arms makers in Damascus, Syria were instrumental in using this material to make swords. The earliest "Damascus swords" date to approximately 500 CE, but they really became very common between 1500–1700. These swords had blades that were extremely tough and strong and represented state-of-the-art metal working technology. Legend has it that the brutal test of a sword made from this material was to behead a slave in a single stroke, and if it could not then it was not worthy of the name "Damascus sword." These blades are really interesting to materials scientists today because no one knows exactly how to make them. No current metallurgical process can reproduce the structure of these blades. A research group at the University of Dresden in Germany examined the blade of a Damascus sword using a high-resolution electron microscope and found that it contained carbon nanotubes [7]. The nanotubes were certainly not added intentionally, but rather they formed "in situ" during the reaction between carbon and iron.

Bulk steel manufacture became possible after the invention of the Bessemer process in 1856, named after Henry Bessemer (he became Sir Henry in 1879). In the Bessemer process a blast of cold air is sent through molten cast iron, in a specially designed vessel called a converter. Once again, conventional wisdom was proved to be wrong. The *cold* air blast actually caused the molten iron to get *hotter* (not colder as might be expected). The reaction between the cold air and the carbon in the cast iron was very rapid, generated lots of energy, and raised the temperature of the melt, so the metal was hotter at the end of the process than at the beginning.

An interesting note on the Bessemer process and intellectual property: the "Bessemer Process" was conceived of independently and almost simultaneously by Henry Bessemer in England and William Kelly, a native of Pittsburgh, Pennsylvania. Kelly had an iron works in Eddyville, Kentucky, that focused on making sugar-kettles for farmers. One day Kelly noticed that a cold air blast could be used to decarburize his iron and if controlled could be used to leave just the right amount of carbon to make steel. Kelly became so excited by his discovery of making steel "without fuel" that his wife thought he had become mentally unbalanced and the other ironmasters rejected him as a crackpot. Kelly skulked off to the woods to build his converters in secret and by 1851 was producing steel by his newly discovered method. In 1856, Kelly learned about Bessemer's work in England and that he has been granted a U.S. patent. Bessemer's patent, number 16,802, was issued on November 11, 1856 (his Great Britain patent was issued the year before). Kelly then filed his own patent and was able to convince the patent office officials of his priority (he was the "first to invent") and he obtained his patent on June 23, 1857—one year after Bessemer. Although William Kelly was declared the first to invent the "Bessemer" converter for producing steel, his success came too late and that same year he declared bankruptcy. His claims were eventually made over to Bessemer, the process became known worldwide as the "Bessemer process," not the "Kelly Process," and within ten years of its invention Bessemer steel was being blown in Britain, the United States, and a handful of European countries.

Bessemer was a prodigious inventor and in his lifetime was awarded a total of 110 patents. For three years I passed beneath a lintel engraved in large stone letters with the name "Bessemer" and walked past a bronze bust of the great innovator marked below with his dates 1813 to 1898. Although Henry Bessemer never attended Imperial College his name is not only memorialized in stone and bronze but also in the Bessemer Laboratory that was completed in 1913 with generous gifts from major businessmen in the iron and steel industries.

Bessemer's original converter converted 365 kg of iron into steel in thirty minutes. The process was soon scaled up to produce 5 tons, then 30 tons, then even more all at the same speed. The molten steel could be poured from the converter to make ingots up to several tons each. This allowed steel to be used for the hulls of ships and large structural joists for buildings. Bulk steel production rose from zero in 1856 to 4.9 million tons in 1900 to 7.3 million tons in 1930. The immediate impact of the Bessemer process was that steel could be produced as cheaply as wrought iron. In fact, the production of wrought iron decreased over that same period and by 1976 it ceased to be produced altogether.

Despite its successes, the Bessemer process had problems. It couldn't get rid of two major contaminants in iron ore: sulfur and phosphorus. These two elements, when present, make steel brittle, leading it to crack easily. Sidney Gilchrist Thomas, a young London police clerk, assisted by his cousin Percy Carlyle Gilchrist, solved the problem in 1875. Their idea, which shocked the established metallurgical community who were not happy at being beaten by outsiders, proposed the use of dolomite for lining the inside of the converter. Dolomite is a natural mineral composed of calcium magnesium carbonate that creates an alkaline or basic environment in the converter as it dissolves in the molten metal. In the melt the dolomite combines with the phosphorus and sulfur forming what is known in the industry as "slag"—simply an impure form of glass. The phosphorus-rich slag has many uses including as a fertilizer for rice cultivation and dry field farming. More recently it has found use as an additive to partially replace Portland cement in the concrete industry.

Thomas and Gilchrist prepared a paper describing the successful application of their invention to be read at the meeting of the Iron and Steel Institute at Paris in 1878. (The meeting was held in Paris rather than London because of the Great Exhibition, which was also taking place that same year in the French capital.) So little importance was attached to the work of the unknown pair that the skeptical and narrow minded British metallurgical community left the paper unread for "lack of time." A Mr. Chaloner who was an attendee at the 1878 meeting wrote the following statement, which Sidney Gilchrist Thomas quoted in his memoirs: "We well remember the sneer as well as 'smile of incredulity,' which spread over that meeting and can testify to the scarcely veiled antagonism exhibited to the unknown youth who had presumed to proclaim the solution to a problem which the leaders of metallurgy had pronounced well nigh insoluble."

The deferred paper of Thomas and Gilchrist was read at the following spring meeting of the Iron and Steel Institute held in London in 1879 to which Mr. Bessemer (not yet Sir Henry) commented: "Whether I should have arrived at the results which the present inventors have arrived at I cannot tell.... I hope and believe they will be able to receive the recompense which their talents and industry deserve."[4] Andrew Holley, an engineer with lots of experience working with Bessemer converters and a colleague of Andrew Carnegie, knew of the work of Thomas and Gilchrist and sent a copy of their paper to mining and metallurgical engineer George W. Maynard, who arranged for a test of the process. The test was successful and when Andrew

Carnegie learned of the result he immediately saw the impact of Thomas's research and acquired for the sum of $250,000 the U.S. license for exclusive use of the technology. Carnegie went on to give Thomas and Gilchrist very high praise indeed: "These two young men, Thomas and Gilchrist of Blaenavon [the location of the ironworks in Wales where they worked], did more for Britain's greatness than all the Kings and Queens put together."

Carnegie is one of the biggest names in the history of steel. He was born in 1835 in Dunfermline, Scotland's historic medieval capital. In 1848 at the age of thirteen Carnegie moved with his parents and younger brother to America. At the age of 24 he was promoted to the position of superintendent with the Pennsylvania Railroad. After a number of successful business ventures, Andrew Carnegie opened his first steel plant in Braddock, Pennsylvania in 1875. In 1883 he purchased the Homestead Steel Works, a rival steel company in Pennsylvania and in 1892 the Carnegie Steel Company was formed. Carnegie recognized the benefits of what we now call "vertical integration": one company owning and controlling all stages of production. Using this approach the Carnegie Steel Company owned everything from the source of iron ore, the coal mines, and the blast furnaces used to produce iron, to the shops where the steel was fabricated into rails for railroads. In doing so Carnegie created the largest steel operation in the world. Just after the turn of the century, Carnegie sold his steel company "and all its holdings" to J.P. Morgan for $480 million (over $14 billion in today's money.) This purchase created U.S. Steel and made Carnegie the richest man in the world. The addition of U.S. Steel to the Dow Jones Industrial Average on April 1, 1901 caused the largest percentage movement due to one stock in the year it was listed. Carnegie spent the rest of his life trying to give the enormous amount of money away. He helped build more than two thousand libraries, Carnegie Hall in New York City, the Carnegie Endowment for International Peace, and Carnegie Mellon University.

In the United States, the driving force for steel production was rails for the great railroad boom that began in the 1840s. Iron rails—normally made of malleable wrought iron—would rapidly wear out under the repeated pounding of heavy trains. An article in the *London Quarterly Review* of 1866 noted that: "On some of the metropolitan lines iron rails, especially if placed on sharp curves, will scarcely last a year.... It thus becomes absolutely necessary to introduce a new material, and that material is to be found in steel" [8].

Even two years after this article was published steel rails were only in use to a limited extent in the United States. But where they were used, it was found that iron rails required replacement seventeen times before the steel rails showed any signs of wear. The steel alloy that was optimum for the manufacture of rails contained 0.82% carbon and 1.7% manganese. The addition of manganese to steel considerably increases its hardness, which is an essential property for durable rails.

Railroads played a very significant role in the development of the United States: they are unfortunately much less important nowadays. The American Industrial Revolution, which began in the northeast of the country required reliable transportation for raw materials, finished products, and workers. On February 28, 1827, the Baltimore & Ohio (B&O) Railroad became the first U.S. railway chartered for commercial transport of passengers and freight. Investors hoped a railroad would allow Baltimore, the second largest U.S. city at the time, to successfully compete with New York for western trade. Baltimore needed a way to reach out to markets in the way that the Erie Canal had done for New York City. The rail line was extended to connect Baltimore with the Ohio River in 1852 and then to Chicago, St. Louis, and Cleveland. Pioneers continued to travel west by covered wagon, but as trains became faster and more frequent, settlements across the continent grew. The first transcontinental railroad was completed in 1869. More miles of track were laid in that single year than any previous one: nearly 5,000 miles.

The impact of steel is in evidence almost everywhere. The Brooklyn Bridge in New York City when it was completed in 1883 was the world's first large steel-wire suspension bridge. But its 5,989 feet span that crosses the East River joining Manhattan and Brooklyn is dwarfed by the Akashi-Kaikyo Bridge in Japan, which at 12,828 feet is currently the world's longest suspension bridge. Although the very earliest suspension bridges used ropes and then iron chains, every major suspension bridge built since the Brooklyn Bridge has relied on steel wire.

The steel I-beam is the key structural component of a skyscraper. The world's first skyscraper, which used a steel skeleton, was William LeBaron Jenney's Home Insurance Building in Chicago that was completed in 1884. It rose to an impressive height of 10 storeys (two more were added in 1890). Although this building is no longer standing it shaped the way that cities are designed. Chicago's Home Insurance Building was demolished in 1931 to make way for another skyscraper, the Field Building (now the LaSalle Bank Building), which stands 45 storeys tall. Skyscrapers allowed cities, like Chicago, to expand upward, rather than outward. It was a concept whose limits New York, and later other major cities, would keep pushing over the following decades.

In 1883 the Naval Appropriations Act allocated $1.3 million to build steel—rather than iron—hulled ships. This was after the success of the USS

Dolphin, which was the first steel-hulled ship in the US Navy. At the time of the 1883 act only one company had constructed any ships with steel plating: Pusey & Jones Company of Wilmington, Delaware. The owner of the company spoke to congress and made the following claim: "vessels built with these sheets of steel, much thinner than we have ever used for iron vessels, and they have been thumped and banged against rocks and stones until one of those boats is all dinged.... Yet there has been no sort of fracture." The House Committee on Naval Affairs was sold on the idea stating that "we have unanimously decided that steel should be used instead of iron." This was the beginning of what the U.S. Naval Institute calls the "steel Navy." Steel is still used today in ships such as the USS *Freedom*, a littoral combat ship used for shallow water missions such as minesweeping and humanitarian relief.

For the past eighty years ships for both military and commercial applications have been produced almost exclusively of welded steel. Early steels did not have the toughness of modern-day alloys. This could lead to catastrophic brittle fracture in cold water a problem experienced by many of the "Liberty ships" built during World War II. Over 2,700 of these cargo ships were built between 1941 and 1945 using a steel alloy that transformed from being ductile when warm to brittle when cold. When the steel was brittle cracks could easily form and then propagate, leading to failure. Since the 1950s there are a range of specialized carbon steels containing manganese and small amounts of the elements vanadium and niobium designed especially for shipbuilding that have been approved by the American Bureau of Shipping.

Steel is usually made using one of two processes: the basic oxygen furnace (BOF) or the electric arc furnace (EAF). About 60% of the world's steel is produced using the BOF: "basic" refers to the magnesia (MgO) furnace linings, which have the same role as dolomite to remove the phosphorus and sulfur. The primary raw materials for the BOF are liquid hot metal from the blast furnace, which makes up about 75% of the charge. The remainder is steel scrap from recycling. Pure oxygen is blown into the furnace at supersonic velocity: greater than 750 mph! The oxygen reacts with the carbon and silicon contained in the hot metal from the blast furnace and this reaction produces an enormous amount of heat that melts the steel scrap creating temperatures of almost 3000°F. A single BOF is 34 feet high, has an outside diameter of 26 feet, and contains about 250 tons of material. The total processing time, called a "heat," for a single batch is about 40 min.

In the United States, only about one-third of the steel is produced using the BOF. Increasingly American steel producers are using the smaller EAF.

These "mini-mills" use an electric arc—a spark between two graphite electrodes—to melt down scrap steel. The temperatures around the arc reach in excess of 2000 °F (1200 °C) and a 100 ton batch of scrap can be melted in 1 h or less. Steel processing has utilized scrap for a long time, but the EAF takes recycling to a new level by using virtually 100% old steel to make new steel. This flexibility allows countries that cannot make iron by the traditional blast furnace route, but have supplies of scrap metal, to make steel using the EAF process. There are currently about 100 operational mini mills in the United States accounting for two thirds of domestic steel production. China dominates the global steel industry producing half of the world's steel.

Steel continues to be the Material of Industry despite numerous challenges from other metals, alloys, composites, and polymers. The form in which steel is used may have changed since the original application, but it remains an indispensable material for the modern world. For example, megatall skyscrapers rely on steel rebar, rather than the I-beam, to reinforce concrete to provide the strengths necessary to reach new heights. Since 2008 the world's tallest skyscraper is the Burj Khalifa in Dubai, United Arab Emirates. It has a total height of 829.8 m (2,722 ft.) and uses 55,000 tons of steel rebar. Its reinforced concrete structure makes it stronger than traditional steel-frame skyscrapers.

According to the US Geological Survey (USGS), the world's reserves of iron ore are more than 800 billion tons, which is equivalent to about 230 billion tons of metallic iron.[5] The major iron ore deposits are in Brazil, China, Russia, and Ukraine. In Russia there is an enormous source of high-grade magnetite known as the Kursk Magnetic Anomaly (KMA). The KMA is located near the city of Magnitogorsk and is estimated to contain 80,000 million tons of iron ore, with an amazingly high metal content (45–65% iron). Plus there are an additional 10 million tons of poorer quality ore. The KMA alone contains enough ore to produce 250 million tons of iron a year for the next 15,000 years! The U.S. has about 110 billion tons of ore containing about 27 billion tons of iron. So while iron may not be as abundant as flint and clay, we are not going to run out of it any time soon.

Iron is the least expensive and most widely used metal today. Iron was critical—in the form of its alloy, steel—in enabling the global transition towards industrialization. As a structural material steel has two major limitations. The first, it competes with lighter materials such as aluminum, plastics, and lightweight composites. About two-thirds of the weight of every car and truck is steel. This includes the body, doors, fenders, and hood. By 2025 the U.S. Government's Corporate Average Fuel Economy (CAFÉ) standards will nearly double the average mileage for light-duty vehicles (cars, sport utility vehicles, and light trucks) to more than 54 miles per gallon. If steel is going to help meet these requirements, then the industry will have to continue development of high-strength lightweight automotive steels.

A team from the Graduate Institute of Ferrous Technology at POSTECH in South Korea published a letter in *Nature* that describes a high-aluminum low-density steel strengthened by the presence of an iron-aluminum compound [9]. The authors claim that the strength and ductility of their new steel improves on those of the lightest and strongest of all metallic materials, titanium alloys coming in at one-tenth of the cost.

The second challenge with steel is the enormous carbon footprint of the iron and steel industry. Almost half of the CO_2 emitted by the entire indus-trial sector comes from four industries; cement, steel, ethylene, and ammonia [10]. (Cement is the subject of Chap. 7 and ethylene is the raw material for the production of polyethylene, described in Chap. 9.) Proposals to reduce the carbon emissions from iron and steel production include the use of hydrogen generated from electrolysis using renewable energy sources rather than coal to reduce iron in a process called direct-reduced iron (DRI). Trials of DRI may begin as soon as 2035 [11].

Notes

1. Data on annual mineral production is available through the US Geological Survey. www.usgs.gov
2. Transition elements occupy the middle portions of the Periodic Table. They are all metals and are characterized as being able to exist in several oxidation states. For example, manganese has oxidation state $+ 2$ in the compound MnO. It is $+ 4$ in MnO_2. Oxygen is always -2.
3. Although we often use the terms strength and toughness interchangeably, they have very specific and different meanings. Strength is the maximum load a spec-imen can withstand when applied either in tension or compression. Toughness is the total amount of work done on the specimen to cause fracture. Toughness is related to strength of a material and also its ductility.
4. Sidney Gilchrist Thomas's story is recounted in his own words in: *Memoir and Letters of Sidney Gilchrist Thomas Inventor* R.W. Burnie (ed) John Murray, London (1891).
5. https://minerals.usgs.gov/minerals/pubs/commodity/iron_&_steel/mcs-2018-feste.pdf Accessed February 2, 2019. The United States Geological Survey has comprehensive data on mineral abundances and uses.

References

1. Hayman, Richard & Horton, Wendy (1999). *Iron Bridge: History & Guide.* Stroud, Gloucestershire: Tempus.
2. Comelli, Daniela, D'orazio, Massimo, Folco, Luigi, El-Halwagy, Mahmud, Frizzi, Tommaso, Alberti, Roberto, Capogrosso, Valentina, Elnaggar, Abdelrazek, Hassan, Hala, Nevin, Austin, Porcelli, Franco, Rashed, Mohamed G, & Valentini, Gianluca (2016). The meteoritic origin of Tutankhamun's iron dagger blade. *Meteoritics & Planetary Science, 51,* 1301–1309.
3. Erb-Satullo, N. (2019). The innovation and adoption of iron in the ancient Near East. *Journal of Archeological Research, 27,* 557–607.
4. Burton, C.V. (1892). A theory concerning the constitution of matter. *Philosophical Magazine, 33,* 191.
5. Williams, Hermann Warner (1935). *A Sixteenth Century German Treatise, Von Stahel Und Eysen : Tr. with Explanatory Notes: Steel and Iron : How to Make Steel and Iron Soft and Tempered . . . Cammerlander at Strasburg, 1539.* Place of publication not identified: publisher not identified.
6. Sullivan, J. W. W. (1951). *The Story of Metals.* Cleveland, Ohio: American Society for Metal and Ames, Iowa: The Iowa State College Press.
7. Reibold, M., Paufler, P., Levin, A. A., Kochmann, W., Pätzke, N., & Meyer, D. C. (2006). Materials: carbon nanotubes in an ancient Damascus sabre. *Nature, 444,* 286. The Damascus sabre that was examined was produced by the famous blacksmith Assad Ullah in the seventeenth century and was housed at the Berne Historical Museum in Switzerland.
8. The London Quarterly Review (1866). *Volume 119* (American Edition, p. 53). New York: Leonard Scott Publishing.
9. Kim, S-H., Kim, H., & Kim, N. J. (2015). Brittle intermetallic compound makes ultrastrong low-density steel with large ductility. *Nature, 518,* 77–79.
10. Gielen, D. J. & Moriguchi, Y. (2003). Technological potentials for CO_2 emission reduction in the global iron and steel industry. *International Journal of Energy Technology and Policy, 1,* 229–249.
11. *The Economist,* Technology Quarterly: towards zero carbon, December 1, 2018, pp. 3–12.

5

Gold—*The Material of Empire*

Gold is a very special material and holds a unique place in human history. It has brought civilizations to the pinnacles of prosperity. The empires of the Sumerians, the Egyptians, the British, the Mesopotamians, and the Romans were built on the power and wealth that comes with the possession of gold. Gold has captured the imagination in song, story, and legend unlike any other material. It has driven nations to war. The Boer War in South Africa was waged because of the British desire to control the then recently discovered gold mines. Unlike many other materials, including those described in this book, gold had—until relatively recently—little inherent technological value. It was not formed into plows that revolutionized agriculture (like bronze). It did not provide the breakthroughs that hastened the Industrial Revolution (like iron and steel). It cannot be used to cut and machine steel that led to mechanization and mass production (like diamond). It was not formed into lenses that revolutionized our understanding of the microscopic and the infinite. Rather, the historical value of gold was based almost solely on its cultural importance: it was, and still is, a symbol of wealth and power. In the words of Sir Thomas More, councilor to King Henry VIII: "They marvel much to hear, that gold, in itself so useless, should be everywhere so much sought, that even men, for whom it was made, and by them hath its value, should be less esteemed." Sir Thomas went on to write: "That a stupid fellow, with no more sense than a log, and as base as he is foolish, should have many wise and good men to serve him because he possesseth a heap of it. And that, should an accident, or a law-quirk (which sometimes produceth as great changes as chance

© The Author(s), under exclusive license to Springer Nature Switzerland AG 2021
M. G. Norton, *Ten Materials That Shaped Our World*,
https://doi.org/10.1007/978-3-030-75213-2_5

herself), pass this wealth from the master to his meanest slave, he would soon become the servant of the other, as if he was an appendage of his wealth, and bound to follow it."[1]

The oldest hoard of gold objects was discovered in Varna, Bulgaria and is dated to the period 4600–4200 BCE. This places it during the Early Bronze Age or Chalcolithic ("copper-stone") period. The approximately 15,000 tiny worked gold items, most of which appear to be for ceremonial use and adornment, were found in 1972 by workers digging a trench for an electric cable. These objects are on permanent exhibition in the Varna Museum of Archeology. The importance of this discovery is that it provides evidence of the advanced nature of the ancient Thracian civilization, which flourished in this part of the Balkan Peninsula. The gold deposits were located at the mouth of the Danube River, which contained one of the largest ancient supplies of the precious metal. Almost all the early gold that has been unearthed was used for religious offerings or as a status symbol. It had a great cultural value in the ancient world and was worn proudly to signify influence and power. There were those who had gold and those who didn't!

Gold has been treasured since ancient times for its beauty and permanence. What is striking about gold, and must have first captivated our ancestors, is its unique yellow color. It is unmistakable. Small nuggets of gold lying in a stream scattered among dull pieces of sand and gravel would have been irresistible. Among all the metallic elements gold is the most resistant to corrosion by the atmosphere. Even after many years of burial, such as in Egyptian tombs, some of the iron artifacts may have rusted and the copper may have turned into a pleasant green due to the formation of an oxide patina, but the gold is as perfect in its appearance as it was on the day that it was hidden away for eternity. This property is often expressed in terms of what chemists call the reduction potential: a measure of how difficult it is for a metal atom to lose an electron and become an ion. Gold has the highest reduction potential, which means it is the most difficult to oxidize.[2]

Gold—chemical symbol Au—can be found at number seventy-nine in the Periodic Table, directly below silver and between neighbors platinum and mercury. Although copper, silver, and gold, which all appear in the same group or column in the Periodic Table of Elements, might be expected to behave in very similar ways they frequently don't, particularly when it comes to their reactivity with oxygen. The explanation is complex because it involves Einstein's theory of relativity.[3] Atoms of gold, silver, and copper all have one outermost electron. In a gold atom this electron is bound much more tightly to the massive nucleus than the outer electrons in silver and copper, which both have much smaller nuclei than gold. Because of the strong attractive forces between the positively charged nucleus and the negatively charged electron the electron has to move at speeds approaching the speed of light to avoid being drawn into the nucleus. Consequently, the outer electron in a gold atom, which would be involved in forming possible bonds with oxygen atoms, is held more tightly and more closely to the nucleus than in silver and copper and therefore is less reactive.

Being in the same group, gold and silver do share some similarities. Both are excellent conductors of electricity and heat. Like its neighbor platinum, gold is a precious metal that is valued because of its inertness. However, platinum is considerably harder than gold and higher temperatures are required to melt it, which makes it much more difficult for jewelers to work with. Although mercury and gold are neighbors, and both heavy metals, they have little else in common. Mercury is the only liquid metal at room temperature, which alone makes it different from all other metals.

Maybe our early explorer picked up the small gold nugget from the streambed and found that it was not hard and brittle like the stones that were nearby, but it could easily be deformed and flattened into a crude disk. Our explorer had discovered one of the superlative properties of gold; its malleability. Gold is the most malleable of all metals, and when it is pure it can be beaten into a leaf so thin you can see through it. Gold leaf is only 0.2 μm (millionths of a meter) thick: about 500 atoms. The Egyptian goldsmiths were extremely skilled craftsmen and produced gold leaf for gilding. Although the Egyptians couldn't get quite as thin as modern techniques allow they did produce very large sheets of gold leaf. An Egyptian coffin in the British Museum in London is gilded with sheets of gold that are almost 9 cm square.

Because pure gold is so soft it is often alloyed with other metals to increase its strength and hardness so that it is suitable for everyday use. When gold is alloyed with silver it is called "electrum," which was an important alloy for making coins as far back as 700 BCE. Common alloying elements include copper, nickel, palladium, and silver. Not only do these additions change the mechanical properties of gold they also impart their own unique colors. White gold is obtained commercially by adding a combination of nickel, copper, and silver to pure gold. Alternatively, but more rarely, a jeweler will make white gold by the addition of palladium and silver. (This combination is rarely used because of the high cost of palladium, a platinum group metal that is essential in the manufacture of electronic circuits.) Yellow gold contains a mixture of copper, silver, and zinc. Adding just copper produces red.

The most unusual of all the colored golds is purple gold, which was first introduced in 2000 in Singapore. The color is due to a compound—a so-called "intermetallic"—that is formed between gold and aluminum. When the two metals are placed in contact and heated they react to form the $AuAl_2$ intermetallic, which is purple [1]. Before its use as Purple Gold™, the gold-aluminum intermetallic was the curse of the electronics industry and poetically known as "purple plague." It formed readily in electronic circuits where gold and aluminum would come into contact at high temperature—for instance when aluminum wires were used to make connections to gold pads on an integrated circuit. Once the problem was identified the industry was able to overcome it by using lower processing temperatures or designing the circuit layout to avoid direct aluminum-to-gold connections. But the unique color sparked the interest of jewelers. Early attempts at making a purple gold for jewelry were unsuccessful because it was not soft enough to be shaped: the gold-aluminum intermetallic is inherently hard and brittle. Aspial Corporation developed a way to overcome this limitation and now purple gold is a commercial product.

Because gold can be alloyed with many different metals it became necessary to develop a system of standardization to quantify the amount of gold that was actually present in a mixture. This is the karat system. The term karat (sometimes "carat" in the United Kingdom) is a measure of the purity of gold. Pure gold is defined as 24 karats. The composition of a gold alloy is then determined by the number of parts of gold it contains: the total being 24. So, 18-karat gold has 18 parts of gold to 6 parts of other metals: a typical 18 karat yellow gold is by weight 75% gold, 14% copper, 9% silver, and 2% zinc. An alternate system uses "fineness": the gold content in parts per thousand. For example, a gold nugget containing 725 parts of pure gold and 275 parts of silver would be referred to as "725-fine."

The karat system is extensively used in classifying gold jewelry, which at 41% represents by far the largest application for gold.[4]

With the abundance of computers controlling everything from our social networks to our refrigerators it is not surprising that after jewelry the electrical and electronics industries are the largest consumers of gold. In third place, almost 20% of gold is used in coinage and currency. In the earliest coins the metals of choice were silver and copper. Gold—or rather electrum—began to be used around 700 BCE in the Lydian-Ionian region of Greece. King Croesus had separate gold and silver coins minted in about 500 BCE. The gold coins were known as "staters"—referring to a standard of weight equivalent to about 8 g. In 300 BCE the Roman Emperor Constantine the Great introduced a new, smaller, gold coin called the solidus, which was 24 karat gold. The solidus remained in circulation for more than one thousand years. The Roman aureus denarius, or "gold penny" was in place from about 50 BCE until 250 CE.

Although gold coins are no longer part of our day-to-day personal transactions—base metals such as nickel, copper, and zinc are used in pennies, nickels, dimes, and quarters—gold in the form of bars (called bullion) and coins are bought and sold every day. The World Gold Council official report of gold holdings shows that at the end of 2020 central banks and other institutions held about 35,000 tons of gold.[5] The United States currently holds over 9,000 tons—the largest gold reserve of any country. A large amount of this gold is stored in the vault of the Fort Knox Bullion Depository near Louisville in Kentucky, Ohio. The vault walls and roof are made of steel and concrete. The vault door weighs more than 20 tons. The first shipment of gold to Fort Knox was in January 1937. The gold is in the form of standard mint bars of almost pure gold or coin bars resulting from the melting of gold coins. The bars are about the size of an ordinary house brick and each contain 400 troy ounces of gold—over 27 lb. In the James Bond movie "Goldfinger," 007 manages to prevent a raid on Fort Knox that is intended to contaminate the entire nation's gold reserve by making it radioactive and therefore worthless. If that happened, at least as the story goes, the US economy, and thereby the world's economy would be destroyed. Bond, of course, saves the day (and the world)!

Gold is a commodity with its price determined by the London Bullion Market. Before the price of gold was allowed to float freely on the exchange markets many countries used the gold standard, where their currency had a value directly linked to gold.

Sir Isaac Newton a recognizable name in the story of gold and arguably one of the most influential scientists who ever lived introduced England into the gold standard, replacing silver "as standard of value for excellence" in 1717 [2]. By that time Newton was already well known for his eponymous laws of motion, the "Newtonian" telescope, and the Universal Law of Gravitation, which described how gravity works. But Newton resigned his prestigious position as a professor at the University of Cambridge to become first Warden and then in 1699 Master of the Royal Mint in London. He set the price at £4.75 per fine ounce of gold, a price that lasted for 200 years. The gold standard was formally adopted by Great Britain in 1816, by Germany in 1871, and by the United States in 1900. In 1933, President Franklin D. Roosevelt introduced the Emergency Banking Act, which banned the export of gold, halted the convertibility of paper dollars into gold, ordered US citizens to hand in all the gold they had, and the next year established a price for gold of

$35 per ounce: a 69% increase from that set in 1900. This value lasted until 1968, when gold was allowed to float freely on the foreign exchange markets. The price of gold then steadily climbed. It reached a peak in January 1980 of $678 per troy ounce and another peak in September 2012 of $1,776. In August 2020 gold hit a record of $1,970.50 per troy ounce.

Possession of gold has been desirable for many individuals and for many countries.

The Egyptians adored gold. It was a valuable and desirable metal to possess because it bestowed power and influence upon the owner. Paintings in Rekhmire's tomb in Egypt dating from 1552–1305 BCE illustrate workers polishing gold and silver vases. But archeological evidence shows that Egyptians were working with gold as far back as 2500 BCE. The desert area around the Red Sea was particularly rich in gold. Egyptian gold also came from Nubia, the "Land of Gold." The Nubian mines contained very large quantities of gold, which was extracted by slaves. The working conditions in the gold mines were horrific. Being a gold miner for the Egyptians might go down as one of the worst jobs in history. Many of the miners worked underground (some of the mines were 300 feet deep) in dingy shafts that were unbearably hot and claustrophobic. The miners risked death from being crushed by falling rocks or from poisoning by the toxic arsenic fumes released from the surrounding quartz rock. Slave labor was cheap and the Egyptians were very thorough miners. They identified, and had their Nubian slaves work, every known accessible deposit to extract every ounce of gold.

An iconic image frequently associated with gold is Egyptian—the facemask of Tutankhamen. Tutankhamen reigned as king of Egypt for ten years from 1333 to 1323 BCE. When he died at the young age of eighteen he was placed in a small tomb together with thousands of objects and treasures that he would need in the afterlife. Many of these were decorated with gold and exquisite jewels. In 1922 when British archeologists Howard Carter and Lord Carnarvon unearthed Tutankhamen's tomb they found elaborate and spectacular examples of gold metalwork. The coffin and facemask were pure gold, and look today exactly the same as they did when they were buried along with the young pharaoh more than 3,000 years ago.

Like the Egyptians, the Romans also had a love affair with gold. They mined gold extensively, initially in Italy but then mainly in Spain, Portugal, and parts of Africa. As their empire expanded, at an increasing pace, to stretch from the Mediterranean to the Black Sea, and from Scotland to Egypt their need for gold increased. Emperors, like earlier leaders acquired gold as a statement of power and supremacy. Wealthy Romans sought gold because it was essential for political influence. Gold was how government expenditures were financed and, importantly, it was used to pay the stipends of the legionnaires: the Roman infantry that enabled the expansion of the empire. The Emperor Domitian who ruled from 81 to 96 gave the first major pay rise for the Roman army. He set the annual stipend of a legionnaire at twelve pieces of gold. In terms of purchasing power that salary is equivalent of about $3,000 today. Not a lot of money, but food, lodging, and "healthcare" were provided. The Roman demand for gold was limitless. Like the Egyptians before them the Romans used slave labor by the thousands in the mines. During the time of Caesar over one hundred thousand slaves were brought in from Gaul to work the mines in Italy. At its peak, gold production is estimated to have been as much as 10 tons of metal per year.

In Europe, gold production declined through the Dark and Middle Ages. Widespread social and economic chaos, incessant warfare, and plagues were the cause. Production only increased again with the Renaissance and the rise of colonialism from the fifteenth century onwards.

The New World provided a rich source of gold and became the destination for western gold seekers. The Spanish conquistadors in the sixteenth century made the acquisition of gold a major part of their conquests. Hernando Cortés uncovered the vast wealth of the Aztecs in Mexico in 1519, which led to a number of other explorations. One of the most notorious was that of Francisco Pizarro who traveled to Peru looking for Inca treasure. Pizarro captured Atahualpa, the Inca ruler of Peru, and ransomed him for enough gold to fill his prison cell. After acquiring eight tons of gold, the lying and treacherous Pizarro had Atahualpa strangled in 1533. The gold captured by the Spanish in Mexico and Peru, and in the later colonization of Brazil, by the Portuguese, led to enormous economic growth in Spain and Portugal, but not so the Central and South American peoples who were conquered and had their resources looted.

England also benefited from this supply of gold, liberating it at the behest of the English crown by piracy. One of the most famous of the privateers was Francis Drake. Known to the Spanish as "El Draqui" (the dragon), Drake was regarded as a hero to the English mainly because of his circumnavigation of the world between 1577 and 1580. The knighting of Drake by Elizabeth I was one of the actions—in addition to the execution of Mary Queen of Scots and Drake's attack on Cadiz in 1587—that led to the Spanish Armada in 1588.

During the Middle Ages in Europe, gold was used to treat everything from mental disorders and syphilis to epilepsy and diarrhea. The early thirteenth century scholar Bartholomaeus Anglicus, in his De proprietatibus rerum ("On the Properties of Things"), described the medicinal benefits of small gold particles: "The filings of gold taken with meat or drink, or as a medicine, are good in leprosy, or at least in effecting a concealment of it, and with the juice of borage and hartshorn, benefits in fainting, and in cardiac passion." The records of famous medieval alchemist Philippus Aureolus Theophrastus Bombastus von Hohenheim (known, thankfully, as Paracelsus) reported a mysterious elixir with almost unbelievably successful curative properties known as "Aurum Potabile" ("drinkable gold")[6]: "Of all Elixirs, Gold is supreme and the most important for us... gold can keep the body indestructible... Drinkable gold will cure all illnesses, it renews and restores." In 1583, alchemist David de Planis-Campy, surgeon to King Louis XIII of France swore by his "elixir of longevity"—an aqueous solution of colloidal gold—for ensuring a long life. As Louis's processor Henry IV of France lived until he was 56 and Louis XIV lived to the ripe old age of 76, the elixir didn't help poor Louis XIII very much. He died at only 41 years of age.

One of the main pursuits of the alchemists was to turn base metals such as lead or copper into gold using the "Philosopher's Stone." Leonardo da Vinci commented on the noble goal of this activity in his *Dell 'Anatomia*, which was published in 1489: "By such study and experiment the old alchemists are seeking to create not the meanest of Nature's products but the most excellent, namely gold, which is begotten of the sun, in as much as it has more resemblance to it than to anything else that is and no created thing is more enduring than gold. It is immune from destruction by fire, which has power over all the rest of created things, reducing them to ashes, glass or smoke."[7]

The search for the stone (or substances that would carry out the transformation) was not successful, but alchemy did make some important contributions to metallurgy. The technique of cupellation, which is used to purify metals, was a result of the work of alchemists. A possibly lesser-known fact about Newton is that he was a passionate believer in alchemy. The famous English economist John Maynard Keynes described Newton as the last of the alchemists: "He was the last of the magicians, the last of the Babylonians and Sumerians, the last great mind which looked out on the visible and intellectual world with the same eyes as those who began to build our intellectual inheritance rather less than 10,000 years ago... [He was] the last wonder-child to whom the Magi could do sincere and appropriate homage" [3]. Alchemy died out in the eighteenth century, but physicists have succeeded where alchemists failed by transforming base metals into gold by bombarding them with highly energetic sub-atomic particles in a particle accelerator. However, this process will never be a commercially viable approach to making gold.

The origin of gold in the universe has been proposed as the result of extremely rare (once every 100,000 years or so) violent collisions between neutron stars.[8] Early in the Earth's history, asteroids containing gold continually bombarded our planet's surface. Over many years, the atoms of gold became concentrated into veins and nuggets. The oldest and largest gold deposits were formed about 3 billion years ago (during the Precambrian era) in the Witwatersrand basin ("the Rand") of South Africa. The gold is present as tiny particles embedded within a pebbly mass of quartz rock: the gold particles striking against the milky white quartz. The Boer miners referred to these rocky structures as "banket" after a popular Dutch almond pastry often enjoyed at Christmastime.

Around 2.5 billion years ago, gold was formed in many ancient mountain ranges such as at Yellowknife in Canada and the Great Boulder Mine near Kalgoorlie, Australia. Here the gold is found in veins—called "lodes"—in quartz rocks. These deposits are newer than those in South Africa. Even younger lodes were formed during the Paleozoic Era, about 300 million years ago. The Californian Mother Lode and the find at Muruntau in Uzbekistan are both from this later period. The Muruntau gold deposit is one of the most significant outside of South Africa with estimated reserves in excess of 5,000 tons.

Even today, deposits of gold are being formed in some active geothermal areas such as Yellowstone National Park in the western United States, parts of Iceland, and in the Taupo volcanic zone of New Zealand. Gold is soluble in the hot water deep within the Earth's crust, where the water often has high concentrations of minerals and salts that help the dissolution process. The gold is then transported to the earth's surface by geysers, hot springs, and mud-pools. Geologists are carefully studying what is happening at these sites to increase their understanding of how gold formed in ancient rocks.

Gold is widely distributed in nature, but it is not abundant like iron and aluminum. On average, the Earth's crust contains about 5 mg (mg) of gold per ton of rock. To put this number into a practical context, a typical wedding band would require all the gold that could be extracted from more than 2,000 tons of gold-bearing rock. Unlike many metals, most gold occurs in the metallic or native state—it is not bound up with oxygen or sulfur in the form of minerals. Gold can be found as small grains or particles and very occasionally large nuggets are unearthed. The largest of all gold nuggets to date is appropriately named the "Welcome Stranger," found in 1869 in Ballarat, which is in the state of Victoria in the southeastern corner of Australia.

Placer deposits, also called alluvial deposits, are relatively young. They were formed less than 100 million years ago when gold-containing rock was worn away by wind and rain in a process called weathering. The gold remained and became concentrated in the beds of rivers and streams. Extracting gold from placer deposits is not difficult; finding it was frequently the challenge. The legend of Jason and the Golden Fleece comes from the use of sheepskin to separate the gold in placer deposits from the associated sand and gravel. The gold particles would be trapped in the fleece and, if enough were present, would impart to it a golden sheen. After the fleece had been left to dry it was shaken or combed to remove the fine gold particles. Most of the gold that was used in ancient times, like that used by the Thracians, came not from the deep, tightly locked, veins but from accessible placer deposits. These sources were easy to find and pea-sized and larger nuggets of gold could be picked out by hand as they become washed from the sand and gravel. The fine particles together with some of the larger nuggets could then be melted into the form of small ingots.

Obtaining the gold that was embedded in quartz rock, the "veins," was much more difficult because of the need for heavy equipment to break the rocks. Agaharchides, a Greek writer, described the process he witnessed that was used by the Egyptians in the second century BCE. First the rock was cracked and broken by heating and then crushed using large hammers. Stone mortars and hand mills further reduced the rock to a fine powder. Lastly, the gold was separated out by washing.

Native gold—although free from oxygen and sulfur—is rarely pure and often contains small amounts of other metals. Differences in the geological formation of the gold ore led to subtle variations in its chemical composition. For example, placer deposits tend to be associated primarily with the metals platinum and palladium. The composition of gold can be used as a characteristic "fingerprint." The "fingerprinting" method, which was established in 1993 by the Anglo American Research Laboratories in South Africa, measures the amounts of impurities that might be present in the gold, including platinum, lead, palladium, thallium, and bismuth. The process works by blasting a small sample of the gold with a neodymium-YAG laser (the same type of laser used for corrective eye surgery). The gold particles vaporize at a temperature of almost 7,000 °C. The individual atoms can be separated—and identified—based on how heavy they are [4]. In South Africa, prosecutors have used chemical composition analysis to trace the origin of stolen gold. The South African mining industry estimates that between 21 and 45 tons of gold, worth nearly $150 million, is stolen each year. This represents between 0.8% and 1.6% of annual global production.

The presence of gold in seawater was discovered in 1872. Seawater contains small amounts of gold—about 0.02 mg per liter. But the volume of the world's oceans is enormous, collectively about 10^{21} liters of seawater, which means there is more than 20 million tons of gold in the world's oceans. At current prices the value of this gold (if it could be extracted!) is almost $400 trillion. One of the proposed approaches to extract seawater gold was that steamers crisscrossing the world's oceans might be able to collect a reasonable quantity by electroplating it onto copper plates suspended in the sea connected with dynamos. However, there is no evidence that this method was ever tried. An earlier proposal was that of the Reverend Prescott F. Jernegan, which he commercialized in 1897 as the Electrolytic Marine Salts Company. To keep the process secret, Jernegan protected his invention, which he called an "accumulator box," as a trade secret—like the recipe for Coca Cola—and refused to patent it. But it is known that the process used electricity and mercury, an element that can easily alloy with gold forming an amalgam from which the gold could subsequently be separated. Having received investment of about $1 million, Jernegan left the United States for Europe in July 1898 under very suspicious circumstances traveling apparently as Louis Sinclair of Chicago. Mr. Sinclair was unknown in Chicago and when captured on a photograph he showed a "striking resemblance" to the good Reverend.[9] Although the science behind the Electrolytic Marine Salts Company's process may have been sound the scheme was not and the company wound up later that year with stockholders losing most of their investment.

The power of gold and the desire to acquire it has been the subject of many myths and legends. One of the best-known is that of Midas, king of Phrygia in Asia Minor, who was given the power by the god Dionysus that everything he touched would turn to gold. The blessing soon became a curse when an elaborate feast became inedible upon Midas' touch. In Nathanial Hawthorn's version of the story even Midas' daughter is turned to gold upon his touch. Midas is cured when he bathes in the river, which washes the curse away. The Sardis River in Turkey, which would have been part of Midas' kingdom, contains large placer deposits that were extracted during ancient times.

Although gold is not abundant, it is widespread. Consequently, several countries experienced major gold rushes. During the second half of the nineteenth century, several major gold finds lead to well-documented gold rushes resulting in the mass migrations of thousands of people to often remote and sparsely populated corners of the world. The most famous of all the gold rushes is the one that occurred in California in 1849. James W. Marshall made the most significant gold discovery at Sutter's Mill, in Coloma, California, on the South Fork of the American River, on January 24, 1848. In Marshall's own description of the event he recounts collecting four or five pieces of gold and testing them to confirm his discovery. Simple field tests for gold include determining its density—it is heavy—and measuring its hardness—it is soft. Four days later Marshall rode to his boss, a farmer named John Augustus Sutter who immigrated to America from Switzerland to make his fortune, to show him the gold. Sutter confirmed Marshall's tests and by the end of 1848, President James Polk mentioned the find in his annual statement to Congress: "The accounts of the abundance of gold in that territory are of such extraordinary character as would scarcely command belief were they not corroborated by authentic reports of officers in the public service."

News of the discovery spread and the California gold rush of 1849 began. Prospectors, known as "forty niners" set off from all parts of the world to seek their fortune in the frontier country in the foothills of the Sierra Nevada mountain range. By the end of 1849 the population of California had more than quadrupled from 26,000 to 115,000. Most of the gold that was found came from placer deposits that could easily be extracted using simple tools such as a pan, a shovel, and a sluice to control the flow of water. The largest single nugget that was found weighed 90 kg and was found at Carson Hill. By 1851, California produced 77 tons of gold, two years later this had increased to 93 tons. By the end of the 1850s the California gold rush was essentially over.

Gold was found in New South Wales in 1823, but it was the discovery of gold in 1851 in the Macquarie River, 200 km west of Sydney, that led to the Australian gold rush. Prospectors came from China, Germany, Great Britain, Poland and even California to Australia. In addition to the finds in New South Wales, gold was discovered in Victoria. The famous Welcome Stranger nugget, mentioned earlier, was found in the state of Victoria. It weighed in at an impressive 78 kg and produced 71 kg of pure gold when melted down (value almost $2.5 million at today's prices). Australian gold rushes led to a peak production of 95 tons in 1856 (at the time worldwide production was 280 tons.) Gold continued to be found in other parts of Australia (essentially following a counterclockwise sequence) and by the end of the nineteenth century Kalgoorlie in Western Australia was the leading gold producing area in Australia. Kalgoorlie has yielded over 1200 tons of gold so far. All the Australian gold rushes were over by the early 1900s and it was not until the mid-1950s that gold production in Australia became significant again.

Gold had been worked in various parts of Africa well before the arrival of European settlers. The richest man in world history, the Mansa Musa of Mali (1280–1337), based his fortune on gold. Although it is impossible to put an accurate number on Mansa Musa's wealth, during his reign Mali produced more gold than any other country in the world: estimates are a ton a year. Tales describe the Mansa's lavish expenses including his pilgrimage to Mecca where he was accompanied by dozens of camels each carrying hundreds of pounds of gold and his army of 200,000 men, which could be paid for with gold.

In West Africa, as in Egypt, the gold that was mined was used for ornamental purposes. There are pictures of Mansa Musa holding a scepter of gold on a throne of gold holding a cup of gold with a golden crown on his head. The dominance of Africa, particularly South Africa, as the world's leading gold producer began in 1886 with the discovery of major deposits in the Witwatersrand. The South African gold rush was in some ways different from those in California and Australia. In South Africa there was already a thriving mining industry based on diamonds. This industry provided an infrastructure of expertise and personnel to work the gold claims, which was particularly useful because it was determined fairly early on that underground mining was going to be necessary to reach much of the gold. South African production peaked in 1970 when 1000 tons of gold was produced. South Africa remained the world's leading producer of gold until 2006 when it was overtaken by China.

The last great gold rush of the nineteenth century was to the Yukon Territory in western Canada, where between 1896 and 1899 the placer deposits by the Klondike River produced some 75 tons of gold (world production at that time was about 400 tons per year). The Klondike gold rush was most significant because of the remoteness of the location, the extremes in the weather, and the considerable hardships involved in accessing the gold-rich areas. But there was no shortage of hopeful—or desperate—miners ready to brave the conditions in search of making their fortune.

Even though the process to obtain placer deposits is very simple, relying only on the density of gold and the force of gravity, it did inflict major environmental blows in those areas where the deposits were mined. Streams and rivers were rerouted causing changes to the local flora and fauna. Downstream effects such as flooding and mudslides became commonplace.

Historically mercury was used in refining gold that had been obtained by panning streambeds. The high toxicity of mercury has severely limited this dangerous practice. Illegal Brazilian miners—called garimpeiros—still use mercury to separate gold they mine in the rainforest of French Guiana. A similar situation is happening in Sudan. But as the Northern African country's economy is in tatters, many feel that they have no other choice than to risk their lives in pursuit of this universally accepted form of wealth.

The process of separating gold from its associated silicate rocks is much more difficult than working with placer deposits and has even greater environmental impacts. The process involves large quantities of energy, a great deal of water, and some very toxic chemicals. First the rock, which arrives as large chunks from the mine, is crushed to a fine powder and put into tanks containing potassium cyanide. Potassium cyanide is a deadly poison and has been used in a number of known murders including the unsolved "Tylenol murders" where eight people died in 1983 in Chicago from random containers of Tylenol that had been spiked with the white cyanide powder. In addition to its known toxicity, a solution of potassium cyanide is one of the few chemicals that will dissolve gold. After the potassium cyanide solution has done its job any undissolved minerals are removed by filtration. By adding zinc to the solution the precious gold is precipitated out. To extract one troy ounce, about 31 g of gold, we need 3 tons of gold-containing rock, 5,000 liters of water, 750 kWh electrical power, explosives, potassium cyanide, compressed air, and 39 man-hours.

According to the United States Geological Survey (USGS) the most recent domestic production of gold was about 250 tons. Gold was produced from lode mines, and both large and small placer deposits (mostly in Alaska). Gold is also produced as a byproduct of processing base metals, chiefly copper. This is what happens at Bingham Canyon Mine near Salt Lake City in Utah in the western United States. The Bingham Canyon Mine, owned by the UK-based company Rio Tinto is more than three-quarters of a mile deep and more than two and three-quarter miles wide making it the largest man-made hole in the world. And it is still growing! The mine is responsible for almost a quarter of all the copper produced in the United States (over 250,000 tons), but also produces 250,000 troy ounces of gold each year, 4 million ounces of silver and 25 million pounds of molybdenum.[10]

The deepest gold mine in the world, Western Deep Levels, is operated at 3582 m below the surface and East Rand Proprietary mines have gone down to 3474 m. At these depths, rock temperatures can reach over 50 °C (125 °F). In large mines more than 70,000 cubic meters of cold air are required per minute to keep the temperature barely tolerable.

The entire operation of obtaining gold is very resource and labor-intensive—it is certainly not environmentally friendly or a low-carbon process. But as a society we want our gold jewelry, gold watches, and gold teeth in addition to the critical industrial applications for gold.

Sir Thomas More described gold as "useless" but the metal has several unique properties that make it essential for many important technological applications. One very useful mechanical property of gold is its ductility. Gold is the most ductile of all metals. A troy ounce—31 g—can be drawn into a wire 40 km in length. Gold wire is critical in many electronic components. Thin gold wires are used to connect the silicon "chip" or integrated

Fig. 5.1 An example of an integrated circuit using gold bonding wire to connect the 'chip' to the package. Gold is also used for the pins and to seal the lid on top of the package. The package itself is made out of aluminum oxide (Al_2O_3)

circuit to the chip carrier, which protects it and allows it to be connected to other circuit components. The use of gold for this application was described in U.S. patent 3,138,743 filed by Jack Kilby of Texas Instruments on February 6, 1959. The importance of this patent goes beyond the use of gold wire. It described the first integrated circuit, which heralded the beginning of the electronics age.

Another advantage of using gold in electronic circuits is that it has one of the highest electrical conductivities of any metal only behind silver and copper. But both of these metals are more likely to oxidize than gold, making it difficult to form reliable electrical connections between circuit components. It is also more difficult to form very thin wires of silver and copper.

Not long after graduating I worked for German precious metal giant Heraeus GmbH, a company based in Hanau near Frankfurt and one of the world's leading manufacturers of gold wire. The finest wire they make is only 15 μm in diameter and has a purity of 99.99%. For comparison, the thickness of a human hair is about 100 μm and the finest copper wire that can be drawn is about 30 μm. Figure 5.1 shows an example of a modern integrated circuit that is housed inside a ceramic package. Everything in the picture that glitters is gold. The individual wires, which are barely visible are about 2 mm long and contain about 20 nanograms of gold! A computer-controlled tool called a wire bonder makes each bond automatically. A single bonder can complete over 20 bonds per second.

NASA's James Webb Space Telescope, due to be launched in French Guiana in October 2021, will search for the first galaxies that formed after the Big Bang.[11] The telescope's eighteen hexagonal mirror segments are made of beryllium—a very lightweight metal—and have been covered with a microscopically thin gold coating. The mirrors make use of gold as a very efficient reflector of infrared light: another superlative property.

The same reflective properties that have made gold invaluable for space exploration have made it a valuable coating in terrestrial settings. A thin coating of gold on windows reflects heat radiation, helping to keep buildings cool in summer and warm in winter, lowering both energy costs and carbon emissions. In Toronto, Canada, all 14,000 windows on the Royal Bank Plaza, the headquarters of the Royal Bank of Canada, are coated in a total of 71 kg of pure gold.

The applications for gold that have been described so far in this chapter are ones that are most familiar because they utilize well known properties of gold. But there is some very interesting research on gold nanoparticles that is changing our perception of this material and opening up new and exciting applications. In fact, gold may turn out to be the single most important material in the rapidly growing field of nanotechnology. Researchers in Japan discovered that when gold is formed into extremely tiny particles, just a few nanometers in diameter, that it has extraordinary and unexpected properties. Not only does it no longer have a gold color it is surprisingly reactive: very different from our traditional view of this material.

In 1982 Dr. Matsutake Haruta, a professor in the Department of Applied Chemistry and the Graduate School of Urban Environmental Sciences at Tokyo Metropolitan University in Japan revealed an unexpected discovery. He demonstrated that gold, the least reactive of our metals, when in the form of nanoparticles 5 nm in diameter or less has extraordinary catalytic properties [5]. In fact, it is an excellent catalyst for a number of important reactions. Catalysts are essential industrial materials. They are used in 90% of the world's chemical processes that manufacture 60% of all chemical products including everything from tires to laundry detergent, pharmaceuticals, and plastics. Although the first application for gold nanoparticles was as "odor eaters" in Japanese rest rooms, a really important potential application is in automobile catalytic converters.

In a paper published in 1987, Professor Haruta showed that in the oxidation of carbon monoxide (CO) to carbon dioxide (CO_2)—an essential function of a car's catalytic converter—that gold worked at lower temperatures than platinum (the currently preferred material) and was more efficient than platinum.[12]

The job of a catalytic converter is to make car exhausts as environmentally friendly as possible. I can remember as a young undergraduate looking down from the Hollywood Hills at the yellowish smog bathing Los Angeles in the late 1970s. The major polluting culprit was the automobile. With the introduction of the Clean Air Act, three-way catalysts (TWC) have been used in emission controls since 1981. An important function of the catalyst is that it converts the very toxic carbon monoxide into the more benign carbon dioxide. The most current version of the TWC uses a metal mixture of platinum and rhodium as the active catalyst—both scarce and very expensive metals. In fact, automobile catalysts are the biggest consumer of the world's platinum resources. Even though gold is a precious metal it is more abundant than platinum and has a lower market price than rhodium. Another very important advantage of gold nanoparticle catalysts is that they can convert carbon monoxide to carbon dioxide at lower temperatures than existing technology.

Considerable amounts of pollutants are emitted from automobile tail pipes within the first five minutes after starting the engine. This is because current exhaust catalysts must be at a temperature of at least 350 °C to function. So for the first part of a commute we are spewing—unless we drive an electric car—raw emissions containing unburnt fuel, carbon monoxide, and nitrogen dioxide into the environment. These so-called cold-start emissions impose a serious pollution problem and require new catalysts that can operate at low temperatures. Gold, in the form of nanoparticles, might be just that catalyst.

Although the applications for gold in the field of nanotechnology currently represent a tiny market and probably in total account for just a few tons—if that—of gold per year, the potential impact of nanoparticle gold is staggering. A material that was once desired only so that it could be hoarded as a demonstration of wealth and power might soon be put to good use to help the environment and transform the chemical industry. Gold nanoparticles are already transforming medicine.

The first use of nanoparticle gold in the field of human medicine is possibly in the traditional Indian Ayurvedic medicine called swarna bhasma (gold ash) that was used from about 2500 BCE in the treatment of a range of diseases including asthma, rheumatoid arthritis, and diabetes. The powdery substance was typically mixed with honey, ghee, or milk and swallowed.

In modern medicine, gold nanoparticles have been used in many applications both to diagnose disease and to treat it. Gold nanoparticles are at the heart of the hundreds of millions of Rapid Diagnostic Tests (RDTs) that are used around the world every year. This well established and critically important technology has changed the face of disease diagnosis in the developing world over the last decade. For example, malaria RDTs work by applying a single drop of the patient's blood to a test strip. The malaria antigens interact with the gold nanoparticles causing a color change on the test strip if malaria is present. The tests are simple, quick (it only takes 20 min), reliable and robust. They can also be used without the need for expensive equipment or highly qualified personnel. According to the World Health Organization (WHO), over 300 million malaria RDTs are sold each year.

Beyond RDT technology other diagnostic technologies exploiting the unique properties of gold nanoparticles are continually being developed. A team at Imperial College in London has a simple diagnostic test for HIV that allows the disease to be spotted earlier and does not require the use of expensive instrumentation [6]. Their approach uses gold nanoparticles, which are formed when hydrogen peroxide reacts with a solution containing gold ions. The target molecule, an enzyme called catalase, if present, can break down the hydrogen peroxide. The speed of the reaction determines the shape of the resulting gold nanoparticles. If the growth is really rapid the gold nanoparticles are spherical, which causes the solution to turn red. When the nanoparticles grow slowly they have more irregular shapes and the solution turns blue. The color change is obvious enough that it can be seen with the naked eye.

A similar approach has been used as an early detection for prostate cancer: the second-most common cancer among men in the United States. The high death toll is partly because the disease has few symptoms in its early stages, meaning that it is difficult to detect.[13] Gold nanoparticles may provide the critical breakthrough.

In addition to detecting disease, gold nanoparticles can be used to treat disease. One example is an injectable compound containing gold that was found to be effective in treating rheumatoid arthritis after World War II. In the twenty-first century, gold nanoparticles are being used extensively in the development of innovative medicines and medical techniques, including needleless vaccine delivery and antimicrobial agents. But perhaps the most promising area of research is in the treatment of cancer.

According to the American Cancer Society, there were over 600,000 cancer deaths in the United States in 2020. This is despite advances in diagnosis and treatment. Treating cancer is made more difficult because often the drugs used in chemotherapy can damage healthy cells. Treatments that target cancer cells directly while limiting the impact on the rest of the body could increase the chance of a full recovery. Research in gold nanoparticles is leading to targeted techniques of delivering drugs and other cancer treatments. CytImmune, a US-based biopharmaceutical company, has developed a method of delivering anti-cancer drugs directly to tumors using gold nanoparticles [7]. The drug is bound to the gold nanoparticles, which are injected into the bloodstream and travel to the site of the tumor, treating it while leaving surrounding tissue largely unaffected. The technology is entering Phase II clinical trials and is undergoing further testing in collaboration with the pharmaceutical company AstraZeneca.

Another company, Nanospectra Biosciences, is taking a different approach to using gold nanoparticles to treat cancer. The company has created "nanoshells," consisting of a gold-coated silica nanoparticle [8]. When these tiny particles are injected into the tumor and then irradiated with a laser of the appropriate frequency they heat up destroying the cancer cells in a process called thermal ablation.

The discovery of the Varna gold horde dating back over 6,000 years illustrates how important gold was to our ancestors. The uniqueness of this metal led to an insatiable and frequently brutal demand for its acquisition. Today there are many applications that rely solely on gold. It has a unique combination of properties that cannot be met by substituting with other materials. While once valued only by what it symbolized, gold is now becoming the workhorse material for nanotechnology finding a wide range of applications in the development of innovative medicines including needleless vaccine delivery and ultrasensitive techniques for early disease diagnostics. While gold is helping in the treatment of diseases such as cancer it has a history of acquisition that has wreaked a great human and environmental toll. Extraction of gold from electronic waste continues our checkered relationship with this indispensable material.

Notes

1. More, Sir Thomas (1516) *Utopia [Riches, Jewels, and Gold]*.
2. Lithium has the lowest reduction potential of all metals, which is one of the reasons it is so effective—and possibly irreplaceable!—as a material in the ubiquitous lithium-ion battery. Lithium is also the lightest of all metallic elements, another advantageous property for a battery material.
3. Egede Christensen, N. (1984). Relativistic band structure calculations. *International Journal of Quantum Chemistry, 25*, 233–261. A detailed paper describing

the electronic structure of the gold atom. The electrons are moving at almost 60% the speed of light.

4. Data from the United States Geological Survey available to usgs.gov.
5. https://www.gold.org/goldhub/data/holders-and-trends. Accessed 5 February 2019. The website of the World Gold Council.
6. *Paracelsus, selected writings*, edited with an introduction by Jolande Jacobi, translated by Norbert Guterman, Pantheon, New York 1951.
7. Leonardo da Vinci (1489). *Dell 'Anatomia*, Milan (trans. E. MacCurdy).
8. Kasen, D, Metzger, B, Barnes, J, Quataert, E, & Ramirez-Ruiz, E. (2017). Origin of the heavy elements in binary neutron-star mergers from a gravitational-wave event. *Nature, 551,* 80–84. This paper describes experimental evidence that supports the origin of gold in the Universe as coming from the violent merger of two neutron stars.
9. Reported as a letter in *The Engineering and Mining Journal*, July 30, 1898 p. 124. Published by The Scientific Publishing Company, New York.
10. https://www.kennecott.com/sites/kennecott.com/files/slide_fact_sheet_fnl4_15_13_315pm.pdf. Accessed 5 February 2019.
11. https://www.jwst.nasa.gov Accessed 5 February 2019. This is the web site for the James Webb Space Telescope. Originally the telescope had been due to launch in late 2018. This, obviously, didn't happen and the new date is Fall 2021.
12. Haruta M, Kobayashi, T, Sano, H, & Yamada, N. (1987). Novel gold catalysts for the oxidation of carbon monoxide at a temperature far below 0 °C. *Chemistry Letters, 16*, 405–8. This has been cited more than 3,200 times. Haruta's definitive review of 1997 (Size- and Support- Dependency in the Catalysis of Gold. *Catalysis Today, 36*, 153–66) has attracted more than 4,800 citations.
13. https://www.cancer.org/content/dam/cancer-org/research/cancer-facts-and-statistics/annual-cancer-facts-and-figures/2018/cancer-facts-and-figures-2018.pdf Accessed 5 February 2019.

References

1. Eakins, D. E., Bahr, D. F., & Norton, M.G. (2004). An in situ TEM study of phase formation in gold-aluminum couples. *Journal of Materials Science, 39*, 165–171.
2. Fay, C. R. (1935). Newton and the gold standard. *The Cambridge Historical Journal, 5*, 109–117.
3. Quoted in White, Michael (1999). *Isaac Newton The Last Sorcerer*. Helix Books.
4. Grigorava, B., Anderson, S., de Bruyn, J., Smith, W., Stülpner, K., & Barzev, A. (1998). The AARL gold fingerprinting technology. *Gold Bulletin, 31*, 26–29.

5. Haruta, M., & Sano, H. (1983). Preparation of highly active composite oxides of silver for hydrogen and carbon monoxide oxidation *Studies in Surface Science and Catalysis* Elsevier. *Amsterdam, 16*, 225–236.
6. de la Rica, R., & Stephens, M. M. (2012). Plasmonic ELISA for the ultrasensitive detection of disease biomarkers with the naked eye. *Nature Nanotechnology, 7*, 821–824.
7. Nilubol, N., Yuan, ZiQiang, Paciotti, G. F., Tamarkin, L., Sanchez, C., Gaskins, K., Freedman, E. M., Cao, S., Zhao, J., Kingston, D. G. I., Libutti, S. K., & Kebebew, E. (2018). Novel dual-action targeted nanomedicine in mice with metastatic thyroid cancer and pancreatic neuroendocrine tumors. *Journal of the National Cancer Institute, 110*, 1019.
8. Rastinehad, Ardeshir R., Anastos, Harry, Wajswol, Ethan, Winoker, Jared S., Sfakianos, John P., Doppalapudi, Sai K., Carrick, Michael R., Knauer, Cynthia J., Taouli, Bachir, Lewis, Sara C., Tewari, Ashutosh K., Schwartz, Jon A., Canfield, Steven E., George, Arvin K., West, Jennifer L., & Halas, Naomi J. (2019). Gold nanoshell-localized photothermal ablation of prostate tumors in a clinical pilot device study. *Proceedings of the National Academy of Sciences, 116*, 18590–18596.

6

Glass—*The Material of Clarity*

Tacoma, a port city in western Washington State, has an important place in the modern history of glass. It is the birthplace of Dale Chihuly, one of the most creative and innovative glass artists in the world. It is also the home of Chihuly's Bridge of Glass, a five-hundred-foot-long pedestrian overpass that links the museum to downtown Tacoma. A few feet above your head as you cross I-705 are over two thousand glass shapes, which, illuminated by natural light, create the sense that you are underwater surrounded by shells, urchins, sea plants, and jellyfish. Figure 6.1 shows one small region of the Bridge of Glass.

Chihuly's glass sculptures have been displayed around the world from Jerusalem to Venice and from Sao Paulo to Bratislava. What makes his work of so much interest to so many people? For me, it is the fluidity of the interaction of the light with the glass. It is never quite the same each time you look at it. The position of the sun, whether it is a clear day or cloudy, the precise location in which you stand—all change how the glass appears. I have visited Tacoma many times and seen the same pieces whether in Union Station or at the Museum of Glass, but each time is like the first time. For an observer of this fluidity, the scale that we see with our naked eyes is a magnification of what is happening at the microscopic level inside the glass.

There are many natural forms of glass, such as obsidian, which was known in Paleolithic times for the sharp edges that could be produced, for example, by striking a piece of obsidian against a large rock. Obsidian is formed when molten volcanic rock solidifies. Trinitite is a natural glass, of sorts, formed

M. G. Norton, *Ten Materials That Shaped Our World,*
https://doi.org/10.1007/978-3-030-75213-2_6

Fig. 6.1 Looking up at the underside of the Chihuly Bridge of Glass in Tacoma, Washington. The bridge connects downtown Tacoma to the Museum of Glass

at the Trinity site in the New Mexican desert where the world's first atomic bomb was exploded on July 16, 1945. It has been suggested, based on the tiny glass droplets that have been discovered there, that the desert sand was carried up into the mushroom cloud, melted at around 8,000 °C (~15,000 °F) and fell as glass rain on the hot sand.[1] In the Badlands of North Dakota buried glass beads provide evidence of an active volcanic region before a devastating asteroid collision with the Earth that occurred over sixty million years ago and coincides with the extinction of non-avian dinosaurs. The power of the asteroid has been estimated at over a billion atomic bombs.

From the examples of natural glass mentioned so far it is clear that glasses may require very harsh—even explosive—conditions to form. One really interesting example of a glass formed under much gentler conditions comes from diatoms, single-celled algae that provided a source of silicon and oxygen, the chemical constituents of flint. The cell walls within each diatom are made of glass. An image of just a single variety of the many thousands of species of diatoms was shown in Fig. 3.1. When the microorganisms die, their glassy skeletons can pile up to form layers thousands of feet thick. Large quantities are sold inexpensively as diatomaceous earth. Diatomaceous earth has many uses from being an effective support for metal catalysts for the Fischer–Tropsch reaction, which is an important process in producing synthetic fuels, to a very absorbent cat litter, an effective bed bug powder, and an ingredient in dynamite.

To make glass in the laboratory, the most straightforward way is to start with sand—ideally one that does not contain much iron. Sand is the essential ingredient of almost all commercially important glasses from that used to make windows and bottles to eyeglasses to the specialized glass for smartphone screens. Sand is also the raw material for the production of silicon, the material of Chap. 11.

The St. Peter Sandstone formation in Missouri is an excellent source of sand because it is so pure (it has a silica content more than 99%).[2] There is an increasing global demand for sand because it is used not only for the manufacture of glass, but it is the biggest ingredient in cement and asphalt. Worldwide mining for high purity sand of the optimum particle size comes at a significant environmental cost.[3] As just one example of this cost, removing sand from the bed of the Yangzi River is believed to have contributed to the extinction of the eponymous dolphin.

Once a suitable source of sand has been found a small quantity is placed into a crucible, typically made of platinum because of its very high melting temperature, and heated in a furnace to a temperature close to 1,700 °C. At this temperature the sand melts transforming into a liquid with a viscosity similar to that of a thick engine oil. Once the crucible is removed from the furnace and allowed to cool, a transparent, colorless piece of glass will be found at the bottom.[4]

The challenge that ancient glassmakers would have had with this method is that they could not get hot enough to melt pure silica—the same problem that early iron workers had when they tried to melt pure iron. What was needed was something to add to the silica to lower the temperature at which a homogenous uniform melt could be produced.

In the clay tablet library of the Assyrian King Ashurbanipal (669–626 BCE) cuneiform texts provide early recipes for making glass.[5] The oldest one calls for 60 parts sand, 180 parts ashes of sea plants, and 5 parts chalk. The ashes of plants—saltwort and glasswort (known as halophytes) were both used—would have provided the ancient glassmaker with a rich source of sodium.

Sodium is an important ingredient in both ancient and modern glasses because it lowers the melting temperature needed to form a glass. High sodium content also makes it easier for the glassmaker to shape the glass into complex shapes. The mixture specified in the ancient cuneiform tablets produces what would now be referred to as a soda-lime silicate (Na_2O-CaO-SiO_2) glass.[6] The ingredients are essentially the same as those used today, but the proportions are somewhat different because of the much higher soda content.

The Roman author Pliny the Elder (23–79 CE) describes the invention of glass in his encyclopedia *Naturalis Historia*: "There is a story that once a ship belonging to some traders in nitrum put in here [the coast of modern Lebanon] and that they scattered along the shore to prepare a meal. Since, however, no stones for supporting their cauldrons were forthcoming, they rested them on lumps on nitrum from their cargo. When these became heated and were completely mingled with the sand on the beach a strange liquid flowed in streams; and this, it is said, was the origin of glass." Nitrum is, like saltwort and glasswort, a naturally occurring source of sodium.[7]

Once a workable formula had been developed to make glass, ancient glassworkers turned their attention to techniques to form their material into different shapes. If you sit and watch glassblowers working in the Hot Shop at the Tacoma Museum of Glass, as I have done many times, you will see the glassblower following a procedure first used by her artisanal ancestors two thousand years ago. First, she takes a metal tube about five feet in length and dips it into a container of molten glass, inside a furnace that is glowing white-hot. It's about 1300 °C in there. The glowing "gob" of glass is about the size of a honeydew melon; it quickly changes color from a bright white to orange as it cools. It has the viscosity of honey on a cool day so the glassblower has to keep it moving to avoid it sagging under its own weight. The glassblower now puts the open end of the pipe to her mouth and blows. The glass gob expands. After only a few minutes the rapidly cooling glass is too tacky to work and it goes back into the furnace to warm up before the shaping continues—back and forth many times—until, eventually, this piece of glass is transformed into a beautiful honeysuckle-yellow basket.

The exact period and location where glassblowing was first developed is not known with certainty, but it happened during the first century BCE, probably in Syria or Iraq, an area with a long association with glassmaking. What we do know is the impact of glassblowing in the ancient world and today. Although this technique was developed over two thousand years ago, the glassblowing pipe has not changed much over time. In fact, the only significant modifications have been automation that allowed mass production of, for example, light bulbs and glass bottles and jars.

Glassblowing created a revolution in the manufacture of hollow glass objects, which the Romans developed to perfection. But the technique was versatile and could be used for more than manufacturing containers; it could

Fig. 6.2 Scanning electron microscope image of the world's smallest vase. This microblown glass structure was formed inside a furnace at 1,800 °C. (The image was recorded by M. J-F. Guinel and M.G. Norton.)

also be used to make flat pieces of glass for mirrors and windows. One process to do this is called the crown method. To make flat glass using this method a glassblower would initially blow the glass to form a large sphere. The hot glass would then be flattened and transferred to a pontil (a solid iron rod rather than the hollow glassblowing pipe), where it is spun to produce a disk that can be as large as 1.5 m in diameter. The center of the disk would contain the raised crown or "bull's eye." Glass panes would be marked and cut from the areas surrounding the bull's eye. This was the process used to make the mirrors in the spectacular Hall of Mirrors in the Palace of Versailles in France, completed in 1684.

For glassblowers to create intricate shapes and complex forms the glass has to be heated so that the viscosity is low enough for it to flow under its own weight. This corresponds to a temperature of around 1000 °C for a typical soda-lime-silicate glass but varies depending on the exact glass formulation. Figure 6.2 shows a very unusual example of "blown" glass that was produced by Maxime Guinel [1]. The microscopic vase is made of silica glass and stands a mere quarter of a millimeter high and has a width less than half that of a human hair. It is resting on a piece of silicon carbide. In the furnace at high temperature, the silicon carbide oxidized to form a layer of silica. The by-product was carbon dioxide. As the gas tried to escape, the glass layer was blown into a bubble that eventually burst when the gas pressure became too great. Because the temperature inside the furnace was 1800 °C the edges of the micro-vase have begun to flow under gravity creating the layered pattern around the mouth. Although we have not found a use for these tiny vases their structures are nonetheless striking.

Glassblowing is not the only technique that has been used to create beautiful objects out of glass. Between 1887 and 1936, the father-son team of Leopold and Rudolf Blaschka produced the world-famous Harvard glass flowers [2]. These realistic models were used to teach botany to students at Harvard University. Together the Blaschkas made an astonishing 850 life-size models representing some 780 species and varieties of plants in 164 families. In addition, they produced over 4,300 detailed models of enlarged flowers and anatomical sections of various floral and vegetative parts of the plants. For the ten-year period July 1, 1890, to July 1, 1900 the Botanical Department at Harvard University signed an exclusive agreement with the Blaschkas for 8,800 Marks (equivalent to about $120,000 today). The Harvard glass flowers are superb examples of lampworking, in which a soda-rich glass is heated in a small flame and shaped by pulling and pinching the softened glass with various tools including tweezers and tungsten picks. The Blaschkas used wire, paint (especially in the early years), varnish, and glue to enhance the models and make them more durable for handling by class after class of students. An example of a Harvard glass flower is shown in Fig. 6.3. This is a plant known as Beardtongue, a part of the Figwort Family, an important source of the valuable heart medicine digitalis. In color and size the model is a perfect facsimile of the original flower in bloom.

What property of glass allows it to be formed into such amazing and complex shapes by glass artists ranging from the Blaschkas to modern glass artists like Dale Chihuly, Karen LaMonte, and Debora Moore? To answer this question, it is necessary to examine the structure of glass.

The critical difference between a glass and a crystalline material, like quartz, gold, or aluminum, is that at the molecular level the glass structure consists of no periodic repeating long-range pattern.[8] In all known crystals, even the most complex structures like proteins, the structure is built on one of seven distinct shapes. These are called the seven crystal systems. The simplest of these shapes—and the most symmetrical—is the cube. We can stack cubes together—like blocks of Lego®—to create an infinite three-dimensional structure. Every component or building block of the structure

Fig. 6.3 An example of a Harvard flower on display at the Harvard Museum of Natural History. (Originally published in C.B. Carter and M.G. Norton, Ceramic Materials: Science and Engineering, 2nd edition (New York: Springer, 2013), p. 491. Republished here under Springer Copyright Transfer Statement.)

is the same—a cube—and the surroundings of every cube in the completed structure are identical. There is structurally no difference between any of the cubes.

There are six more complicated crystals systems than the cube, but they all share the same requirement that they can be stacked to create a three-dimensional structure with no gaps.

In the glass structure, there is no such repeating structural unit. If you were to examine a piece of glass using a very high-powered microscope (one that doesn't yet exist!) you would find that everywhere you looked the structure would be different. It would not be possible to find two areas that were identical. In that way, at least, the structure of a glass resembles that of a liquid like water. The difference, at room temperature, is that water has a very low viscosity—it flows with no appreciable resistance to adopt the shape of its container—whereas glass is highly viscous. In fact, the viscosity is so high that at normal temperatures glass behaves as though it were a solid.

It is important to acknowledge that we don't actually know what the structure of glass is, because we have no way of "seeing" it. The model that we use to visualize glass structure was proposed by Norwegian-American physicist William Houlder Zachariasen in the 1930s.[9] The basic idea is that certain oxides (the most frequently used example is silica, SiO_2) form a random three-dimensional network when they are cooled slowly from a melt. The structure has no repeating pattern that would be found in, for instance, a crystal of quartz even though they are chemically identical. If certain oxides (most commonly, soda, Na_2O or lime, CaO) are melted together with silica they will break the glass network, which weakens the structure because there are now more broken bonds making the glass easier to melt and shape. Sodium is the most effective of these network-modifying oxides that is why it is such an important component in ancient glasses and modern commercial glasses.

Window glass contains about 14% by weight of soda: more than any other oxide except silica. Over 16% soda is used in the glass formulation for lamp bulbs. The reason for the difference in composition is that the two products are formed by very different process and most window glass spends half of its life outside.

Experimental research over many years has shown that Zachariasen's so-called Random Network Model of glass structure is not wrong, but experiments cannot confirm that it is correct! What we do know is that glasses are very much more complicated than crystals. Using the Random Network Model, the glass industry has developed formulations that we use for everything from windows to wine glasses and from laboratory beakers to beer bottles.

Colored glass has been produced since ancient times. Egyptian glassmakers as early as 3000 BCE were making glass for jewelry and producing small delicate glass containers to hold scented oils and small tubes that were popular containers for the iconic black eye make-up known as kohl worn by Egyptian queens and noble women. Jewelry glass was often made to match the colors of prized stones such as garnet, peridot, and the bright blue of lapis lazuli. When glass is intentionally colored it is often done so by the addition of metal oxides to the formulation. One of the best-known colorants is cobalt oxide, which forms a bright blue glass called "cobalt blue." The cobalt atoms are thought to join the silica network randomly replacing silicon atoms in the structure rather than breaking bonds in the way that sodium does.

Possibly the most famous example of cobalt blue glass is the Portland vase, which is housed in the British Museum in London. This vase was made in Rome during the first century and consists of dark blue glass decorated with white figures. During the eighteenth century the Dowager Duchess of Portland (who died in 1785) acquired the vase, hence its name. The British Museum purchased the vase from the seventh duke of Portland in 1945.

On the other hand, most bright yellow, orange, and red colors in glasses are not the result of adding metal oxides but rather are due to the presence of colloids (nanoparticles) formed within the glass. The most striking of these colors is the deep ruby red produced by the presence of colloidal gold. Gold has been used since Roman times to produce red glass. The Lycurgus cup—also in the British Museum in London—is a particularly beautiful and striking example. The 16-cm high cup was made in Rome during the fourth century. The decoration illustrates the story of Lycurgus, the ill-fated king of the Thracian Edoni, who was strangled by vines after taunting the god Dionysus. The color of the glass is particularly unusual because it appears a deep red when viewed in transmitted light. In reflected light, it is pea green. This effect is known as dichromatism (meaning two colors) and is the result of both gold and silver colloids being present within the glass. If the glass contained colloids of only pure gold, the glass would not be dichromatic. It would be monochromatic—red.

The process of making ruby glass, one of many technological accomplishments of the Romans, was lost in Europe for over one thousand years until it was rediscovered in Germany in the seventeenth century. The color is widely referred to as the "Purple of Cassius" after German physician Andreas Cassius (the younger) who has frequently been credited with its rediscovery in 1685 following the publication of his work *De Auro* in Hamburg. However, L.B. Hunt from the Johnson Matthey Company in London has recounted the true story of this remarkable color.[10] More than a quarter century before Cassius's publication, German chemist Johann Glauber had described the preparation of colloidal gold and between 1679 and 1689 ruby glass was being produced in large quantities at the Potsdam glass factory under the direction of Johann Kunckel. But the naming of the color has remained the "purple of Cassius" in honor of Dr. Cassius.

During the Victorian era a lighter more delicately colored pink glass was very popular. This glass is called Cranberry glass and contains less gold than the darker more intense ruby red glass. Cranberry glass was often used for bowls, lamps, and vases.

Forming gold colloids is not the only way to produce red glass. In the early 1890s glass chemist Nicholas Kopp discovered that selenium could be used to make an intense red glass. In 1926 Kopp set up his own glass works and used "selenium red" glass for the production of railroad and traffic signal stoplights. When cadmium is added to the glass batch as cadmium sulfide it is possible to produce colors ranging from yellow to red depending on the relative amounts of sulfur and selenium. Cadmium sulfide produces a yellow glass. A sulfur to selenium ratio of 3 to 1 yields orange, whereas a ratio of 2 sulfurs to 3 selenium atoms gives a red color.

A drawback with using cadmium sulfide and cadmium selenide to color glass and glazes is that the colors are unstable under the typical glass and ceramic processing conditions. One way to stabilize the color is to trap it inside a stable clear transparent zircon shell. Zircon is a crystalline oxide of silicon and zirconium and has been used since 1986 to produce encapsulated or protected pigments. Nowadays, these structures would be called core-shell nanoparticles.

Ancient glassmakers transformed the vision of what was possible with glass by producing increasingly creative and beautiful shapes with rich colors. But one challenge glassmakers faced for centuries was how to make clear, colorless glass.

Typically, most glass produced in antiquity had at best a green or yellow tinge because of the presence of dissolved iron impurities. This is perhaps not surprising, as iron is one of the most abundant metals in the Earth's crust and would be present in most sands. The color is due to iron atoms directly entering the glass structure much in the same way that cobalt does to produce cobalt blue. It is very different from the process in which small colored particles, much like pigments in paint or magnetite in obsidian, cause color.[11] For the eye of the perfectionist, the greenish-brown color was undesirable.

From about 70 CE decolorized glass was being produced, the result of a series of trial-and-error additions to the glass formulation and the use of higher-purity raw materials. However, the problem was not completely satisfactorily addressed until the Venetians in the middle of the fifteenth century used the mineral pyrolusite, a naturally occurring form of manganese dioxide (MnO_2). Pyrolusite is a black crystalline mineral that looks very similar to obsidian. Although it is not a glass, pyrolusite has the same shiny glassy appearance.

To understand how pyrolusite can produce a clear, colorless glass it is necessary to appreciate an important property of transition metals: many can be readily oxidized and reduced by losing or gaining electrons, respectively. Manganese is a transition metal. It is next to iron in the first row of the Periodic Table of Elements with one fewer electron. When small amounts of pyrolusite are added to the glass melt the manganese reacts with the iron. The iron, which is mainly present in the troublesome ferrous state (giving the green color) is oxidized by losing an electron to form the less problematic ferric state, which produces a pale yellow. At the same time the manganese is reduced to a colorless form by acquiring an electron. One electron is trans-ferred from the iron to the manganese. The resulting glass resembled in appearance the much-prized rock crystal—a natural form of quartz—and was called *cristallo* by the Venetians. Because of its unmatched clarity, cristallo was the most superior form of glass that had ever been produced up to that time.

However, the pyrolusite remedy is not permanent. Over time, as the glass is exposed to sunlight it develops a distinct shade of violet due to the gradual oxidation of the manganese. (This is the same source of the purple color seen in potassium permanganate crystals that are used in the water treatment industry for purification of wastewater.) Some old windows, particularly in Belgium and the Netherlands, have a noticeably purple tinge due to many years of sun exposure.[12] The glass is sometimes referred to as "sun-purpled" or "sun colored amethyst".

It is interesting to pause briefly and consider the state of the field of chem-istry in the fifteenth century when the Venetians were systematically using reduction and oxidation (so called "redox") reactions to decolorize glass. At that time, the element oxygen had yet to be discovered. That discovery, which would be made by Joseph Priestley, was still 300 years away. Consequently, the term "oxidation" had not been coined. Although iron was one of the elements of alchemy, the element manganese was not isolated until 1774 by Johan Gottlieb Gahn and was one of the initial sixty-three elements placed in Mendeleev's Periodic Table of the Elements published in 1869. So, the basic scientific understanding behind using redox chemistry for decolorizing glass was completely absent from the minds of the Venetian glassmakers, yet they had the ability to systematically modify the glass formulation in such a way that they could produce the most remarkable glass.

Another very important addition to glass, which produced a new kind of brilliance and clarity, was lead. During the closing years of the seventeenth century English glassmaker George Ravenscroft was experimenting with silica glass formulations that contained red lead (Pb_3O_4). What he created was a unique English glass that became known as "lead crystal" glass, a rival to the famed Venetian cristallo. As mentioned in Chap. 3, Chinese ceramists had used lead to make glazes as far back as the Warring States period (475–221 BCE.)

Lead was explored as an alternative to pyrolusite because it too helped to neutralize the undesirable green–brown color. The additional oxygen, whether the lead is added as red lead or litharge (PbO), oxidized the iron to the less obvious pale yellow. But what made the addition of lead so impor-tant was that because lead is such a heavy element it significantly increases the refractive index of the glass. To the beholder the presence of lead made the glass sparkle, particularly when it was cut and polished, like diamond. (This was the same benefit potters obtained using lead glazes.)

Refractive index is a measure of the degree that light is bent as it travels into, or out of, a transparent material. It is simply a ratio of the speed of light travelling through the material to the speed of light travelling through a vacuum.[13] Regular window glass has a refractive index of about 1.5. Adding lead increases the refractive index of glass to over 2. (For comparison, diamond has a refractive index of 2.4.[14])

The lead content in Ravenscroft's lead crystal glass has been determined to be about 15%. Nowadays lead crystal glasses contain anywhere from 18 to 38% lead oxide. For tableware to be marketed and sold as "lead crystal" the lead oxide content must be greater than 24%.[15] The British glass industry, in particular, expanded rapidly following the success of lead crystal glass and during the eighteenth century it achieved the leading position that it held for a hundred years.

The beautiful drinking glasses of this period are collectors items. Examples of Ravenscroft's lead crystal glass are in the British Museum in London, the Toledo Museum of Art, the Muzeum Narodowe in Warsaw, Poland and major museums in cities around the world.

The development of clear, colorless glass and improvements in the production of flat glass between the fifteenth and the eighteenth centuries provided a material basis for the extensive use of mirrors. The most notable example is the magnificent Hall of Mirrors the most famous room in the Royal Palace of Versailles just outside Paris. Each of the 357 mirrors is built up of small polished sections of cristallo glass.

But in some cases, maybe clarity was not required. Hanging in the National Gallery, just off Trafalgar Square in London is a painting by Spanish master Diego Velázquez. The Rokeby Venus as it is known was painted between 1647 and 1651; a little over twenty-five years before the commencement of the Hall of Mirrors. It shows a nude Venus, the Goddess of Love and the most beautiful of goddesses, looking at her reflection in a mirror that is held by her son Cupid. The painting is a remarkable work of art and Venus and Cupid are sharply defined against a black sheet and a red curtain. The only part that lacks clarity is the distorted, dull image of Venus in the mirror. Maybe Velázquez was not familiar with the best quality mirrors of the time or is Venus's true identity being hidden. Possibly the blurred image of Venus is Velázquez contrasting the illusion of art with the reality of life.

Although seventeenth century glass was far from free of internal distortions often the result of stresses induced during cooling from the shaping process, once it was possible to make clear glass free from the presence of coloring impurities, glass opened up the beauty of the microscopic and the magisterial. In 1665, Robert Hooke, a curate's son from the Isle of Wight, published *Micrographia*,[16] a book that Samuel Pepys called "the most ingenious book that ever I read in my life." What Hooke had done was use a microscope to magnify tiny objects, which he then replicated by detailed drawings in his book. At the time there was no way to photograph the objects under the microscope. Hooke's drawing of the head of a grey drone fly showed, for the first time ever, the fly's amazingly complex compound eye with its array of hundreds of tiny ommatidia. Knowing this structure then allows us to appreciate why it is so difficult to catch a fly—it has a large angle view of the world and can detect fast movements.

Hooke's microscope used three lenses, which Hooke described in *Micrographia* as: "The Microscope, which for the most part I made use of, ... was contriv'd with three Glasses; a small Object Glass a thinner Eye Glass, and a very deep one: This I made use of only when I had occasion to see much of an Object at once; the middle Glass conveying a very great company of radiating Pencils, which would go another way, and throwing them upon the deep Eye Glass. But when ever I had occasion to examine the small parts of a Body more accurately, I took out the middle Glass, and only made use of one Eye Glass with the Object Glass." Hooke's accomplishments would have been groundbreaking even with the most perfect of modern-day glasses lenses. But the lenses he used, although state-of-the-art, were far from perfect and had significant aberrations, which produced distorted images. As Hooke himself noted, to get more accurate observations he actually removed a lens to reduce the total distortion of the image. The lower magnification produced a clearer image.

For materials scientists, one of Hooke's most important observations was when he examined broken pieces of Cornish flint. Inside the exposed cavity he found "a very pretty candied substance," which on further examination proved to be "a multitude of little *Crystaline* (sic) or *Adamantine* bodies, so curiously shap'd, that it afforded a not unpleasing object." This structure, first observed by Hooke, explains why as we saw in Chap. 2 flint exhibits the special type of conchoidal fracture surface creating sharp edges that were such an effective tool for our earliest Stone Age ancestors.

The regularity of the quartz crystals with their sharply defined faces led Hooke to the conclusion that they result from a regular arrangement of particles. He was correct, and these regularly arranged particles turned out to be atoms. In 1849, almost two hundred years after the publication of *Micrographia*, August Bravais presented his ideas on crystallography to the French Academy of Sciences where he showed that all possible three-dimensional crystals are of fourteen distinct types [3]. Each point in these fourteen lattices represents one or more atoms in the actual crystal. These fourteen lattices are known as Bravais Lattices.

Hooke showed abundant curiosity about the natural world and his study of crystals extended even to looking at those formed from dried urine that he found sticking to the sides of a urinal. The plate-like crystals were a mixture of rhomboids, squares, and rectangles.

Observing these crystals Hooke proposed a procedure to deal with painful and troubling kidney stones. Recent images of kidney stones obtained by a team at the University of Chicago using a powerful scanning electron microscope show that Hooke's observations were perfectly correct in identifying the shape of the stones [4].

Maybe the most spectacular revelation to come from Hooke's microscope was that fossils are the remains of once living creatures. In the middle of the seventeenth century, fossils were thought to be deformed fragments of rock that had fortuitously formed these life-like shapes. The microscopic evidence suggested otherwise and Hooke noted that what we now call ammonites must be "the shells of certain Shellfishes, which, either by some Deluge, Inundation, earthquake, or some such other means, came to be thrown to that place, and there to be filled with some kind of mud or clay, or petrifying water." In a lecture at Gresham College in London, Hooke made the profound inference: "parts which have been sea are now land," and "mountains have been turned into plains, and plains into mountains, and the like."

In 1609 Italian physicist and astronomer Galileo Galilei became the first person to point a telescope skyward. Although that telescope was small and the images fuzzy, Galileo was able to make out mountains and craters on the moon's surface, which he carefully rendered in ink drawings. Galileo observed a ribbon of diffuse light arching across the sky—which would later be identified as our Milky Way galaxy—and found it was filled with stars packed so densely they appeared as a cloud. Although the ancients had observed with their naked eye the five planets in the sky named after the Roman gods—Mercury, Venus, Mars, Jupiter, and Saturn—Galileo's telescope revealed detailed information about the planets and their orbits.

Galileo's first telescope was a simple arrangement of two glass lenses placed inside a hollow cylinder. One lens, the objective, was convex the other, the eyepiece, was concave. The magnification power was only eight times, but later versions increased this to ten and eventually to a thirty-power telescope that when turned toward Jupiter revealed its four largest moons—Io, Ganymede, Europa, and Callisto.

The most incredible technological achievement in obtaining clear, defect free glass of unparalleled clarity must surely be the glass optical fiber. The idea of sending optical signals along hair-like glass fibers was first proposed in 1966 by Charles Kao and George Hockham researchers at Standard Telecommunication Laboratories (STL) in the United Kingdom.[17] Light is transmitted from one end of an optical fiber to the other by a process called total internal reflection. Assuming that the light is not absorbed or scattered by defects in the glass it will be repeatedly reflected from one internal surface to another along the length of the fiber. Kao and Hockham determined that for glass fibers to be used to transmit optical communications the attenuation—or loss—must be less than 20 decibels per kilometer (dB/km). In other words, 1% of the incident light must remain in the fiber after travelling 1 km. At the time of Kao and Hockham's proposal fibers with losses of 1,000 dB/km or even higher were common.

In order to make glass perfect enough to transmit light over long distances a new processing method was needed. Rather than forming glass in the traditional way by mixing powders together, melting them, and then cooling the hot liquid, the glass for an optical fiber is made by reacting ultra-high purity gases together and growing the glass inside a hollow silica tube. The process is known as modified chemical vapor deposition (MCVD) and was developed by the famous Bell Laboratories in the United States. The process starts with a purity silica glass rod known as the preform. The preform is able to be rotated while it is heated. Reactant gases, silicon tetrachloride ($SiCl_4$) and germanium tetrachloride ($GeCl_4$), are fed in at one end of the preform. Inside the hot tube they react with high-purity oxygen to form a glass layer that deposits on the inside walls of the preform. About 1 g of glass is formed per minute. The composition of the glass—the ratio of silicon to germanium—can be modified through the thickness of the glass layer, which produces a gradual change in refractive index allowing the path of the light to be carefully controlled to decrease losses.

When a sufficient glass thickness has been built up, the tube is collapsed by heating to almost 2000 °C. Glass fibers are drawn from the collapsed preform at a rate of more than 10 m per second. The resulting fiber consists of a high-purity silica glass cladding with a core having a layered glass composition to maximize total internal reflection.

Just over twenty years after Kao and Hockham's publication, optical fibers were crossing the Atlantic Ocean and another ten years later the Pacific Ocean. By the end of the twentieth century much of the world's telecommunications was travelling through clear glass fiber optic cables.

Glass opened up a world too small to see with the naked eye and a universe too large to imagine. Hooke may have seen crystal shapes that suggested ordered structures, but his microscope could never have unraveled the structure of glass, a structure that still eludes us today. In the middle of the seventeenth century when Hooke was staring at images created by his microscope he would have been challenged to imagine that in the twenty-first century that it would be through our interactions with glass that we communicate and connect with the outside world. As we tap, swipe, spread and pinch the smartglass screens of our smartphones we are interacting with glass in a new and unique way.[18]

Notes

1. The Trinity site is open to visitors twice a year—the first Saturdays in April and October—and although it is now illegal to remove any of the remaining Trinitite, it can be purchased. https://www.wsmr.army.mil/Trinity/Pages/Home. aspx Accessed 19 February 2019. Web site of the White Sands Missile Range, which gives information about the Trinity site.
2. https://dnr.mo.gov/geology/geosrv/imac/stpetersandstone.htm Accessed 19 February 2019. Website of the Missouri Department of Natural Resources with extensive information on the St. Peter Sandstone formation and its industrial importance.
3. For cement the size of the sand particles is particularly important. Desert sand from the United Arab Emirates, for example is too fine to be made into cement. River sand that has been shaped by the flow of water is coarser and more suited to the application.
4. Silica is known as a glass-forming oxide because it readily forms a glass even when cooled very slowly from its melting temperature. Other glass formers include germanium oxide, boron oxide, and phosphorus oxide. Glasses are not just oxides. Metallic glasses can be formed, but the cooling rates are very high. In some cases more than a million degrees per second.
5. https://www.britishmuseum.org/research/research_projects/all_current_proj ects/ashurbanipal_library_phase_1.aspx Accessed 19 February 2019. The oldest surviving royal library in the world is that of Ashurbanipal, King of Assyria (668-around 630 BCE). The library consists of more than 30,000 cuneiform tablets and fragments.
6. The glass industry uses a number of terms to refer to different metal oxides: soda is sodium oxide (Na_2O), lime is calcium oxide (CaO), silica is silicon oxide (SiO_2), alumina is aluminum oxide (Al_2O_3), litharge is lead oxide (PbO), potash is potassium oxide (K_2O).

7. Rasmussen, S.C (2012) *How Glass Changes the World*, Springer Briefs in History of Chemistry, 85 p.12. Gives an explanation of the different forms of "nitrum" that might be encountered. The word 'nitre,' which most recently refers to sodium nitrate, has only acquired that meaning within recent centuries. Originally it meant carbonated alkali, something that effervesced with vinegar or other acid, and when dissolved in water was a cleaning agent. The ancient Egyptians obtained native soda called 'nitrike' from lakes such as those in Nitria The Greek word became 'nitron' and in turn became the latin 'nitrum' and the European 'nitre.' Thus the Greek 'nitron' used by Hippocrates in the fifth century BCE, the Latin 'nitrum' of Pliny in the first century CE, and their English equivalent 'nitre,' all apply to the soda obtained from either evaporitic lakes or plant ash.

8. We say that glass possesses only short-range order; crystals have long-range order.

9. Zachariasen, W.H. (1932) The atomic arrangement in glass *Journal of the American Ceramic Society 54*, 3841–3851. Zachariasen's model was consistent with experimental observations made around the same time using X-ray diffraction from glass obtained by B.E. Warren at Massachusetts Institute of Technology, e.g., Warren, B.E. (1934). The diffraction of X-rays in glass *Phys. Rev. 45*, 657–661.

10. Hunt, L.B. (1976) The true story of purple of Cassius: The birth of gold-based glass and enamel colours Springer. The article can be downloaded from link.springer.com. The author L.B. Hunt was with Johnson Matthey & Co. Limited, London. When I worked at Cookson Group, Johnson Matthey was one of our main competitors for ceramic pigments. The mechanism that yielded the striking red color was unknown until it was explained by Austrian-Hungarian chemist Richard Adolf Zsigmondy, winner of the 1925 Nobel Prize in Chemistry.

11. Naturally occurring glass is almost always colored because it contains large amounts of impurities. The green color of Trinitite comes from ubiquitous iron impurities present in the sand. Obsidian is most often black or very dark brown because of the presence of iron in the form of tiny particles of magnetite, a black magnetic oxide of iron. Green and red obsidian can also be found, but these are not as common. Red obsidian gets its color from the presence of microscopic hematite crystals. Hematite is a naturally red oxide of iron that is an important source of iron for steel making. It was also used, before the advent of digital music and streamed movies, for magnetic recording onto videocassette tapes. The green form of obsidian also owes its color to iron, but in this case the iron is present in particles of the silicate mineral feldspar. Even though in all three cases the color of obsidian is due to the presence of iron, the form in which the iron is present in the glass makes an enormous difference.

12. Gaffield, Thomas (1872) The action of sunlight on colourless and coloured glass *Report of the Forty-Second Meeting of the British Association for the Advancement of Science* pp. 37–8. The paper describes the author's experience with discoloration of glass plates placed on the windowsills and roof of his house in Boston. The month of June was particularly harsh on the glass and caused discoloration in a few days.

13. We observe the effect of refractive index when we look at a "bent" spoon in a glass of water. The spoon is, of course, not bent but the light coming from the submerged part of the spoon is reaching our eye from a different angle than the light coming from the handle.

14. Although diamond is comprised of light atoms (carbon) the open structure of diamond with its very strong carbon-to-carbon bonds are the physical factors that give diamond a high refractive index.

15. The Sheffield Assay Office provides a definition of what constitutes lead crystal glass. A glass containing more than 24% lead oxide is "lead crystal." If it is less than 24% it is called "crystal glass.".

16. Hooke, Robert (1665). *Micrographia, or some physiological descriptions of minute bodies made by magnifying glasses, with observations and inquiries thereupon*, London. The first important work on microscopy. Not only does it provide detailed observations of the microscopic, but it also provides a glimpse of some of the challenges Hooke faced working with his subjects. For example, the ant (p. 203) proved to be particularly problematic and Hooke had to sedate him with 'Brandy,' which 'knock'd him down dead drunk, so that he became moveless.' The ant recovered an hour later and 'ran away.'.

17. Charles Kao was awarded the Nobel Prize for Physics in 2009 for his discovery of how light can be transmitted through fiber-optic cables. The classic paper by Kao and Hockham is Kao, K.C and Hockham, G.A. (1966). Dielectric-fibre surface waveguides for optical frequencies *Proc. IEE. 113*, 1151–1158.

18. According to the Pew Research Center 77% of Americans own a smartphone. By 2020 it is estimated there will be 6 billion smartphones in circulation. http://www.pewinternet.org/fact-sheet/mobile/ Accessed 21 February 2019.

References

1. Guinel, M.J-F. & Norton, M.G. (2005). Blowing of silica microforms on silicon carbide. *Journal of Non-Crystalline Solids, 351*, 251–257.
2. Evans Schultes, Richard & Davis, William A. (1982). *The Glass Flowers at Harvard*. Cambridge: Harvard Museum of Natural History.

3. Bravais, A. (1850). Mémoire sur les systèmes formés par des points distribués regulièrement sur un plan ou dans l'espace. *J. Ecole. Polytech., 19*, 1–128.

4. Arvans, D., Jung, Y.-C., Antonopoulos, D., Koval, J., Granja, I., Bashir, M., Karrar, E., Roy-, J., Musch, M., Asplin, J., Chang, E., & Hassan, H. (2017). *Oxalobacter formigenes*–Derived bioactive factors stimulate oxalate transport by intestinal epithelial cells. *Journal of the American Society of Nephrology, 28*, 876–887.

7

Cement—*The Material of Grandeur*

The isle of Portland juts into the English Channel just south of the seaside resort of Weymouth. Joseph Aspdin was born almost 250 miles north east of Portland in the Hunslet district of Leeds. But the most widely used construction material in the world connects their names forever. On October 21, 1824, Aspdin was granted British Patent GB5022, titled "An Improvement in the Mode of Producing an Artificial Stone." The patent states that this artificial stone is a cement, to be called Portland cement, which is to be employed for stuccoing houses and water cisterns, and for other uses. I don't know if Joseph Aspdin ever visited the Isle of Portland—maybe he did on a summer vacation with his wife Mary, but he was certainly well aware of the high-quality stone quarried in this picturesque south coast location. Portland stone had been used in a number of important and highly visible buildings including the iconic St. Paul's Cathedral in London that was designed by Sir Christopher Wren and completed in 1708. It was the resemblance between this Jurassic building stone and the appearance of the set mortar that led Aspdin to coin the name "Portland cement" for his invention.

The process to form Portland cement is described as follows: "puddle" or powder from roads laid with limestone, or the limestone itself used for that purpose (when the other materials cannot be procured in sufficient quantities), should be calcined, and when slaked should be mixed with a "specific" quantity of clay and water to an "impalpable" state, by manual labor or by machinery, and should then be put in a "slip pan", and be dried by the sun, or by fire flues beneath the pan, until entirely deprived of the water. The

M. G. Norton, *Ten Materials That Shaped Our World*,
https://doi.org/10.1007/978-3-030-75213-2_7

whole is then to be broken into lumps, and again calcined in a lime kiln, after which it is to be reduced to powder by grinding, rolling, or pounding, when it will be fit for use.[1] What Aspdin had invented was a mixture of limestone and clay, which formed a powder that, when mixed with water, could be used as a fast-setting, low-strength mortar. About two thousand years before Aspdin's invention, the Romans found that if they mixed volcanic ash called pozzolana—a mixture of silica (SiO_2) and alumina (Al_2O_3)—from Mount Vesuvius that buried the ancient city of Pompeii in 79 CE with lime they formed a cement that resisted both fire and water. When the cement was blended with aggregate; that is, stones, sand, or volcanic rocks such as pumice, it made a concrete that was hard as stone. Concrete was the material that allowed the Romans to create a mighty city built to last for the duration of their empire.

Three incredible examples of the Romans' use of concrete are the Roman Baths, the Pantheon, and the Coliseum. Built by Agrippa in 27 BCE, the Pantheon was entirely reconstructed by Hadrian in its original present form in 125 CE. The most remarkable feature that is immediately apparent when you enter the Pantheon is its enormous dome: 142 feet in diameter. It is the largest unreinforced concrete dome ever built. At its peak, some 140 feet above the floor is a 27-foot hole, an oculus, which provides the only source of light. On a sunny day the light streams in, illuminating with decreasing intensity, each of the five increasingly large concentric circles that form the interior of the dome until it eventually reaches the marble-tiled floor. At certain times of day the light fills the incredible structure with an intensity impossible to match with even the most modern lighting systems. This vision must have acted as an inspiration for many Romans and for those that followed. It continues to inspire to this day.

American artist Michael Schultheis uses a representation of the Pantheon in his series of paintings that explore the geometry of Greek mathematician Archimedes. The artist describes the relationship between his artwork and the Pantheon: "Archimedes explored the elegant relationship between the volume and surface area of a sphere. The Pantheon in Rome was constructed in the shape of a hemispherical dome on top of a circular cylinder so that the interior sphere that contains the dome rests of the center of the marble floor. This geometry can be seen while standing inside the Pantheon and in my paintings it pictorially serves as the dioramic space in which I travel into the work. The Oculus at the top of the Pantheon can be seen at the top of these paintings, from which sunlight and the genius of Archimedes stream in." The same representation of the Pantheon can be seen in the later paintings by Schultheis in the series *Dreams of Pythagoras: Pythagorean Accolades* is shown in Fig. 7.1.

Fig. 7.1 *Pythagorean Accolades* (2014) Acrylic on canvas. A gift from the artist, Michael Schultheis. The painting is on display at Elmina White Honors Hall, Washington State University, Pullman WA

The illumination begins at the top of the painting, where the outline of the dome of the Pantheon is visible, then spreads to the entire canvas.

The Pantheon is the one Roman monument that has survived to the present day almost intact and owes its survival both to its imaginative design and to the fact that it was converted into a church in 609 CE. The exterior foundation walls are made of pozzolana cement over a layer of dense stone aggregate.[2] Stresses caused primarily by settling and movement over the past two thousand years, and by the occasional earthquake, are relieved by the formation of cracks. There was no steel reinforcement at that time to support tensile stresses and Roman pozzolanic cement, just like modern Portland cement, is much weaker in tension than in compression.

However, the genius of the Roman architects was such that they designed the wall structure so that cracks would form where their impact on the integrity of the structure would be minimal. Cracks caused by tension are in regions where the overall forces are predominantly compressive—playing to the inherent strength of concrete—so that they would tend not to propagate and lead to structural failure [1]. Stresses that might have razed other buildings have not been able to topple the Pantheon.

Although the dome of the Pantheon remains as a remarkable achievement of Roman imagination and engineering skill, possibly their most important contribution to building was the arch: a structure made of brick reinforced with concrete. Once the concrete had hardened, the design of the Roman arch directed all the stresses down toward the pillars: there were very little lateral forces, which could weaken the structure eventually causing it to fail. Roman arches were to be seen in their most dramatic form in the building of aqueducts, a design much admired by the Victorians, who used it extensively—although for supporting railways rather than transporting water. The brick-and-concrete arch is also much in evidence in Rome's mighty Coliseum, the exterior of which comprises three orders of eighty arches.

The great accomplishments of the Romans in developing concrete technology were lost after the fall of the Roman Empire in 476 CE and not rediscovered until almost a thousand years later in 1414 with the discovery of manuscripts of the Roman architect and military engineer Marcus Pollio Vitruvius describing the cement-making process using pozzolana. The lost manuscripts were found in a Swiss monastery and subsequently printed in 1486, which made it easier for others to imitate his work.

The next important step in the history of concrete happened in the late eighteenth century when English engineer John Smeaton discovered a process for producing a quick drying hydraulic lime, a lime that hardens underwater like Roman pozzolana. Smeaton's process involved firing limestone-containing clay until it turned into clinker: a glassy-looking lumpy residue. The clinker was then ground to produce a powder. Smeaton used his cement in the historic rebuilding of the Eddystone Lighthouse in Cornwall, England. The reef on which it stands was the source of many—indeed hundreds!—of shipwrecks. Smeaton's lighthouse was completed in 1759. After 126 years in service protecting commercial shipping in the area, erosion of the rock upon which the lighthouse stood meant that a replacement was required. (The hydraulic lime had stood the test of time. It was the surrounding rocks that had given way.) No time was lost in building another—the fourth—lighthouse on the dangerous Eddystone Rocks. Douglass's Tower was completed in 1882 and proudly stands to the present day.

The modern era of cement technology began with the awarding of Aspdin's patent for Portland cement. Portland cement is a hydraulic cement like pozzolana, but has a well defined composition with the oxides of calcium (CaO), silicon (SiO_2), aluminum (Al_2O_3), and iron (Fe_2O_3) making up 90% of the material. The potential and versatility of this new product was immediately recognized, and within a year it was being used in major engineering projects. For instance, Sir Marc Isambard Brunel used Portland cement in the construction of the Thames Tunnel, which connects south-of-the-river Rotherhithe with Wapping on the north bank. The Thames Tunnel was the first tunnel under a navigable river anywhere in the world. When it opened in 1843—eighteen years after the project broke ground—it was described as the Eighth Wonder of the World.[3] On the first day, fifty thousand people descended the staircase and paid a penny to walk through the tunnel. By the end of the first three months a million people, equivalent to half the population of London, had come to walk the length of the 1,300 feet tunnel 75 feet below the river. The Thames Tunnel quickly became the most successful visitor attraction in the world.

But why was the tunnel—or any tunnel—needed? The reason was the daily passage of three thousand tall-masted ships that would use the Thames to transport cargo and passengers to and from London, which at the beginning of tunnel construction had overtaken Beijing as the largest city in the world. A tunnel was the only way to get cargo across the river without halting the passage of the ships. The challenge was that no one had successfully tunneled across a river before.

The technology to enable such an undertaking existed. It had been developed by Sir Marc Isambard Brunel and patented by him in 1818 as The Shield: British Patent GB4204. The tunneling shield used by Brunel and his then 19-year-old son Isambard Kingdom allowed miners to dig while inside a multi-story protective frame. Bricklayers would follow behind building the supporting wall as the diggers advanced. The entire 1,300 feet tunnel was dug out in four-inch-wide strips using only short handled spades. The working conditions for the miners were appalling. They would be showered in raw sewage and had to dodge flames from ignited methane gas. After a four-hour shift the men were physically exhausted and close to collapse. Flooding was a further danger. The Tunnel flooded five times. In the worst flood six men drowned and the younger Brunel barely escaped with his life.

When the tunnel finally opened it was only for pedestrian traffic. The first person to use the tunnel and travel under the Thames was Henry Brunel—Isambard's son—who was carried through as a baby by his proud father. Unfortunately, the project had run out of money and could not afford to build the huge double helix ramps necessary to get horses and carts down into the tunnel. In 1865, twenty-two years after the tunnel had been completed, the East London Railway Company purchased the tunnel with the intention of linking it to the national railway network: allowing a north–south connection through London under the Thames. Four years later, in 1869, trains started to run through the tunnel. For the first time this engineering marvel was doing what it was intended to do—carrying freight across the river.

The East London Railway was electrified in 1913 and eventually incorporated into the London Underground as the East London Line, making the Thames Tunnel the oldest tunnel in the oldest underground system in the world. The Thames Tunnel—enabled by Portland cement—is the original Tube!

America got its first concrete street in 1891 when George W. Bartholemew, founder of the Buckeye Portland Cement Company, Main Street in Bellefontaine, Ohio. Bartholemew had to post a $5,000 bond to guarantee his work for five years. After the success of Main Street other neighboring streets were paved in concrete. One of those streets was Court Avenue, which was paved in 1893 and exists today as the oldest concrete pavement in service in the U.S. The total amount of paving on four streets surrounding the Logan County Courthouse in Bellefontaine was 7,700 square yards at a cost of $2.25 per square yard. (For comparison, a square yard of concrete today is about $100.)

By 1897 Sears Roebuck was selling 50-gallon drums of imported Portland cement for $3.40 each. At this time there were more than 90 different formulas for the "same" product, so consistency became a problem. In 1917 the National Bureau of Standards (now the National Institute of Standards and Technology) and the American Society for Testing and Materials (now ASTM International) established a standard, ASTM C150, defining the compositions of ten types of Portland cements based on their application. For instance, Type I is a general cement. Type III is for applications where early high strength is required.[4]

The chief ingredients in Joseph Aspdin's Portland cement and Roman pozzolanic cement are di- and tricalcium silicates and tricalcium aluminate. These compounds are often represented using the abbreviated forms C_2S, C_3S, and C_3A, respectively. We know from considerable experience that this mixture hardens in the presence of water. What we don't understand very well is exactly how the mixture gets hard. The late Dr. Hamlin Jennings, an expert on cement chemistry and former head of the Concrete Sustainability Hub at the Massachusetts Institute of Technology said: "I don't think there is any building material used in the world today that is more poorly understood than Portland cement."[5] His comments could probably be broadened to say that we know less about cement than we do about any other widely used material.

Jennings spent thirty-five years trying to understand the complex reactions that occur as a wet cement paste transforms into a hard-wearing long-lasting structural material [2]. At a very simplistic level we can identify a series of reactions that occur. The first reaction involves the tricalcium aluminate or C_3A reacting with water. This reaction is very rapid and occurs within the first few hours after the concrete has been poured. There are two important products of this reaction: needle-shaped crystals of ettringite and a great deal of heat. The chemical formula for ettringite can be written using our abbreviated notation as C_3AH_6, where H represents water (H_2O). As the cement begins to set, the ettringite needles grow into rods and interlock, creating a rigid structure. The second reaction—hardening—is slower: it begins after about 10 h and takes more than 100 days to complete. The primary product of the hardening reaction is a gel called tobermorite (written as $C_3S_2H_3$), which surrounds the cement particles and bonds everything together. The hardening reaction also produces large quantities of heat. So, the structure of hardening cement consists of a hot mixture of interpenetrating ettringite rods held together with tobermorite gel.

Even days after the hardening process has begun there are still pores in the material. These pores act as "weak links" in the concrete, and it is critical that as hardening continues the number of pores and their size decreases. The hardening reactions actually continue for decades! Roman author and naturalist Pliny the Elder wrote in his book *Naturalis Historia* around 79 CE that "as soon as it [concrete] comes into contact with the waves of the sea and is submerged, becomes a single stone mass, impregnable to the waves and every day stronger." In 1995 samples were taken from the Hoover Dam, which is located at the Nevada-Arizona border in the southwestern United States and was completed in 1935. Tests showed that the hardening reactions were still occurring sixty years later. The concrete was continuing to get stronger. More recent studies using a wide range of analytical instrumentation by Marie Jackson a professor in the Department of Geology and Geophysics at the University of Utah and published in the journal *American Mineralogist* have shown that reactions in Roman pozzolanic cement have actually been going on for thousands of years [3].

Both setting and hardening reactions generate an enormous amount of heat over the first few hours and days after the initial pour. In very large concrete structures heat generation represents a potential problem if left unchecked. If the surface of the concrete cools quickly, while the internal temperature is still high stresses can build up that may lead to cracking. Cooling pipes must be embedded in the concrete to remove the heat. In the case of the Hoover Dam, the construction consisted of a series of individual concrete columns rather than a single block of concrete. It is estimated that if the dam were to have been built in one single continuous pour, it would have taken 125 years to cool to the surrounding air temperature. The resulting stresses created as the concrete cooled would certainly have caused the dam to crack and eventually crumble. The cooling pipes were left in place, filled with grout, and help to reinforce the concrete structure.

The Grand Coulee Dam in Washington, completed in 1942, is the largest concrete structure in the United States and contains 12 million cubic yards of concrete. (For comparison the Hoover Dam was completed after pouring a little over 3 million cubic yards.) Placing of concrete for the Grand Coulee Dam used the same methods employed for the construction of the Hoover Dam. Cold water from the Columbia River was pumped through pipes embedded in the setting concrete, reducing the temperature in the columns from 105 °F to a cool 45 °F. The resulting contraction caused the dam to shrink about 8 inches in length: the gaps were filled with grout.

Construction of the Grand Coulee Dam created thousands of jobs during the Great Depression. Electricity produced by the dam's enormous generators provided the power to produce lightweight and tough aluminum alloys, which were critical for building the planes, ships, tanks, to fight World War II. The proximity of Hanford, in south-central Washington, close to the Grand Coulee Dam was a major consideration in choosing this site for the production of plutonium, which was used in the first atomic bomb detonated in the New Mexico desert near Alamogordo, and the bomb (called Fat Man) that was detonated over Nagasaki, Japan on August 9, 1945.

Following the war the power generation capacity of the dam was increased, which allowed the economic development of the Pacific Northwest. The Grand Coulee Dam provides three-quarters of the entire electrical power demand of the region. Over the course of a year an average of 21 billion kWh of electrical energy are supplied to Washington, Oregon, Idaho, Montana, California, Wyoming, Colorado, New Mexico, Nevada, Utah, Arizona, and Canada. The economy of my home state—Washington—developed around fishing and logging industries during the nineteenth century. Access to abundant and affordable energy allowed expansion of the industrial base, which now includes manufacturing and tech giants including Boeing, Amazon, and Microsoft.[6]

Until 2006 the Grand Coulee Dam enjoyed the recognition of being the largest concrete structure in the world. That title now goes to the Three Gorges Dam that spans the Yangtze River by the Chinese town of Sandouping in Hubei Province. The Three Gorges Dam is 1.4 miles long, at 607 feet tall is five times the height of the Hoover Dam, and incorporates 37 million cubic yards of concrete. The dam has the capacity to generate more than 22 gigaWatts (GW) of electricity: the largest installed capacity of any dam in the world.

While concrete is very strong under compression, as evidenced by the persistence of the dome of the Pantheon and the arches of the Coliseum, it has a low resistance to tensile forces. The reason is that pores such as those present because of incomplete hardening reactions or small cracks formed by stresses generated during cooling are weak points that can lead to the formation of catastrophic cracks. Whereas compressive forces push the pores or cracks together, closing them and stopping them from growing, tensile forces have exactly the opposite effect and cause any pre-existing cracks to rapidly expand. In 1951 Swedish engineer Waloddi Weibull published a statistical model that could be used to assign a probability that a component would fail based on the theory that a material will fail at its weakest point. In brittle materials like concrete the weakest link would correspond to the largest crack; the biggest flaw [4].

To increase the design flexibility and widen the applications for concrete it was necessary to overcome its inherent tensile weakness. A successful approach was developed for plant pots but would become the technology that would lead to the largest structures ever created: to reinforce concrete with iron. Iron, like most metals is equally strong in tension and compression.

Early attempts to reinforce masonry using iron had been unsuccessful because the metal would rust causing the bond to the surrounding masonry to weaken. This problem was overcome with the use of concrete, which is virtually waterproof once hardened and was an ideal host to prevent corrosion of the iron. The resulting material—reinforced concrete—could now resist both compressive and tensile forces in almost equal measure.

To demonstrate the versatility of reinforced concrete, Joseph-Louis Lambot, a French engineer, built a rowboat in 1848 made of concrete that was reinforced with a mesh of iron bars. The boat was exhibited at the Exposition Universelle in Paris in 1855 and apparently aroused considerable interest. Lambot patented reinforced concrete as an alternative to wood in buildings, but it seems that he did not develop his ideas any further. Another Frenchman, Joseph Monier, was more persistent.

Monier was a gardener and in 1849 he made tubs for orange trees out of concrete using a material very similar to that used by Lambot for his rowboat. Monier took out several patents to protect his invention. In 1867 he was awarded a patent for the construction of plant tubs and the like by means of concrete reinforced with a meshwork or rods or wires. But his most important place in history was French patent 120,989, "Sleepers and blocks of cement and iron for common roads and railways," issued in 1877. This patent covered reinforced concrete beams—an indispensable modern building material.

The first home built using reinforced concrete was a two-story servant's cottage constructed in England by William Boutland Wilkinson in 1854. Wilkinson was a plasterer from Newcastle-upon-Tyne in the northeast of England. He used concrete reinforced with iron bars and wire rope for the floor and roof of the cottage. What is particularly interesting about Wilkinson's design was that he arranged the metal reinforcements in such a way that they shouldered the tensile stresses. This, together with Wilkinson's correspondence shows that he had an excellent understanding of the engineering principles of reinforced concrete.[7]

Between 1850 and 1880 François Coignet used Portland cement reinforced with steel rods to construct several large houses in both France and England. Initially the steel rods were used to prevent spreading of the exterior walls. Later on they were used as structural components increasing the strength of the concrete just as Wilkinson had done. In 1875 American mechanical engineer William E. Ward completed the first reinforced concrete home in the United States. Known as Ward's Castle, it still stands on Magnolia Drive in Port Chester, New York. Concrete was used in the construction rather than wood because Ward's mother was afraid of fire. The Great Chicago Fire had occurred only 4 years previously and killed 300 people and left more than 100,000 people homeless. In order to make the house more socially acceptable the exterior was designed to resemble masonry; Coignet's buildings were disguised likewise. Ward's house became listed on the National Register of Historic Places in 1976 and from 1976 until 1992 it housed the Museum of Cartoon Art.

During the late nineteenth century, the use of steel-reinforced concrete was being further developed more or less simultaneously by a German, Gustav Adolf Wayss, a Frenchman François Hennebique, and an Englishman in America, Ernest L. Ransome. Hennebique started building steel-reinforced concrete homes in France in the late 1870s. By 1900, through a network of dozens of agencies across the world, he had constructed nearly 3,000 buildings. One example of these buildings is the seven-story Weaver Flour Mill in Swansea, Wales, which was completed in 1897.

The Ransome Engineering Company was founded in California in 1870 under Ransome's leadership and used a patented system—US Patent 305,226, issued in 1884—involving twisted square rods, rather than the straight rods used in earlier versions of reinforced concrete. The twisted roads had greater contact area with the concrete creating an improved bond between the two materials, which would resist the metal rods being stretched and drawn. Ransome was sent to San Francisco by his father and became one of the most important pioneers of reinforced concrete in the United States and was responsible for buildings across the country. In 1889 he built the Alvord Lake Bridge in the Golden Gate Park in San Francisco. This was the first reinforced concrete bridge in North America and among the first three or four in the world. Now almost 150 years later this bridge and some of San Francisco's original reinforced concrete sidewalks still stand as a testament to the quality of Ransome's material and the durability of the construction.

In 1879, Frankfurt engineer Wayss bought the rights to Monier's patented technology and used it for something much larger and more impressive than plant pots. In 1904 Wayss's firm built the Isar River Bridge at Grünewald, Germany, which at 230 feet (69 m) became the longest reinforced concrete span in the world. Wayss's firm commissioned Berlin civil engineer Matthias Koenen to investigate the theoretical basis of reinforced concrete. Koenen quantified what Wilkinson and others had realized. In his memoirs Koenen writes: "I was resolved to give the matter my full attention because it was completely clear to me that I now had the basic conditions for a new type of construction before me; for besides the concrete body in compression I saw not only the possibility of ties and walls, but primarily also a chance of building elements to resist bending; all one had to do was to increase the insufficient tensile strength of the concrete considerably by placing a suitable number of iron bars in the tension zone which through the known resistance to sliding on the concrete during deflection and the formation of the moment of resistance would have to be called upon to contribute" [5].

At the turn of the twentieth century French architect Auguste Perret, with his brothers Gustave and Claude, designed and built an apartment building at 25 Rue Franklin in Paris using steel-reinforced concrete for the columns, beams, and floor slabs. The building had an elegant façade, which helped make concrete more socially acceptable. 25 Rue Franklin also launched reinforced concrete as a critical design element in modernist architecture. In Perret's words: "One must never allow into a building any element destined solely for ornament, but rather turn to ornament all the parts necessary for its support."

The building was widely admired and concrete became increasingly considered as both an architectural feature as well as a building material. Kenneth Frampton and Yukio Futagawa praise 25 Rue Franklin: "This apartment building with which Perret established his reputation is to be regarded as one of the canonical works of twentieth-century architecture" [6].

One year after Perret's "brilliant use of reinforced concrete" the first concrete high-rise building—The Ingalls Building—was completed in Cincinnati, Ohio. Melville Ingalls, for whom the building is named, spent two years convincing city officials to issue him the necessary permits. Skepticism was high, because the height record for a concrete building stood at seven storys: the Weaver Flour Mill in Wales. The Ingalls Building would be more than double that. Melville Ingalls prevailed and his vision for a concrete skyscraper was realized. The Ingalls Building stands 16 storeys— 210 feet tall—and is still in use today. The success of the Ingalls Building led to the acceptance of reinforced concrete for high-rise construction in the United States.

In the years following completion of the Ingalls Building in 1904 reinforced concrete was used in several major building triumphs including Terminal Station in Atlanta, Georgia, and the Marlborough Hotel in Atlantic City, New Jersey. These projects established reinforced concrete as a worthy competitor to steel. However, most high-rise buildings continued to be made of steel. That situation changed dramatically in 1962 with Bertrand Goldberg's construction of the 60-story twin towers of the Marina City complex on the banks of the Chicago River. These corncob-shaped towers became the tallest residential buildings in the world and the tallest reinforced concrete structures.[8]

A year after the Ingalls Building was completed in Cincinnati two states west in Oak Park, Illinois the Unity Church burned to the ground. Unitarian Universalist minister Rodney Johonnot wanted its replacement to be a modern building that would embody the principles of "unity, truth, beauty, simplicity, freedom, and reason." The person awarded the commission to realize Johonnot's vision was Frank Lloyd Wright. Wright's design broke almost every existing convention for traditional Western ecclesiastic architecture. On the choice of construction material Wright states: "There was only one material to choose—as church funds were $45,000. Concrete was cheap."[9]

In terms of design, Wright envisioned a building consisting of a series of arrayed and stacked concrete forms whose raw material was exposed:

uncovered. In September of 1909, Wright's masterpiece was dedicated and is regarded as the greatest public building of the architect's Chicago years. Because its unique design bore little resemblance to the other churches in the neighborhood, it was decided to rename it Unity Temple. While Wright had chosen reinforced concrete for its economy the completed building ultimately cost nearly twice the contracted price due to complications encountered during construction.

Nothing illustrates the strength of reinforced concrete and its versatility more than the Sydney Opera House: a building that in the words of US architect Frank Gehry "changed the image of an entire country" and shaped the course of twentieth century architecture. Synonymous with inspiration, imagination, and grandeur the Sydney Opera House is one of the most recognizable buildings in the world. In 1956 the New South Wales Premier, The Hon. Joseph Cahill, announced an international competition for the design of an opera house for Sydney. More than 230 entries were submitted from around the world. The winner was a relatively unknown thirty-eight-year-old Dane, Jørn Utzon. His vision for a sculptural, curved building on the Sydney Harbor broke radically with the cube and rectangular shapes of modernist architecture. Construction was expected to take four years: It took fourteen. The Sydney Opera House was opened by Queen Elizabeth II on October 20, 1973 and quickly became Australia's number one tourist attraction welcoming more than eight million people every year. On June 28, 2007, the Sydney Opera House was included on the UNESCO World Heritage List placing it alongside the Taj Mahal, the ancient Pyramids of Egypt and the Great Wall of China as one of the most outstanding places on Earth. The International Council Report on Monuments and Sites to the World Heritage Committee stated: "[The] Sydney Opera House stands by itself as one of the indisputable masterpieces of human creativity, not only in the twentieth century but in the history of humankind.[10]"

The striking structure of the Sydney Opera House was made possible through the use of the "thin-shell" technique, which was pioneered by Spanish architects and engineers Eduardo Torroja y Miret and Felix Candela Outeriño. In this technique very thin (just a few inches), but extremely strong, shells of reinforced concrete are used. The small amount of material means that thin-shell structures are unusually economical with a corresponding decrease in their carbon footprint.

In 1930, the Spanish engineer Eduardo Torroja designed a low-rise dome for the market in the port city of Algeciras in the extreme south of Spain. The dome spans 150 feet, with a thickness of only 3½ inches. Felix Candela, a Spanish mathematician-engineer-architect who emigrated to Mexico in 1939 and practiced mostly in Mexico City, was probably the most accomplished person when it came to building using the thin-shell techniques. In 1950 he came to international attention for his design of the Cosmic Ray Laboratory at the National Autonomous University of Mexico in Mexico City. At its thinnest point the roof is only 5/8 inch thick and varies up to a maximum of 2 inches. Other famous thin-shell structures built by Candela in Mexico City are the Church of La Virgin Milagrosa in 1955, which has a warped roof of reinforced concrete 1½ inches thick and the Church of San Vicente de Paul five years later.

The ease with which concrete can be shaped and its durability have inspired its use in many structures that shape amd define our modern world. From the innovative, literally ground-breaking work, of Monier, Ransome, Wayss, Perret and others in the late nineteenth and early twentieth centuries concrete has today become a multi-billion-dollar industry that employs more than 2 million people in the United States alone. Worldwide, the production of cement exceeds 4 billion tons per year making it a global $170 billion-a-year industry. The only substance people use more of than concrete, in total volume, is water.

Construction of hydroelectric dams—enabled by concrete—is taking place in many countries across the globe including Brazil, China, Ethiopia, Tanzania, and Vietnam. The goal with each of these massive construction projects is to provide those countries with a sustainable source of energy. Unfortunately, dams create their own environmental problems. In the construction of the Three Gorges Dam in China 1.4 million people were displaced, a habitat that supports 300 species of fish and the endangered Chinese River Dolphin (Baiji Dolphin) was disrupted, and the destruction of many towns and villages occurred. Assessing the pros and cons of hydroelectric power is not easy and the discussion will continue probably without resolution. But within the necessary energy mix for the twenty-first century many countries are considering that hydroelectricity must be a major component.

Production of concrete is responsible for 5% of the world's anthropogenic carbon dioxide emissions. In the United States, only fossil fuel consumption (for transportation and electricity generation) and the iron and steel industry release more of the damaging greenhouse gas. The process for making cement calls for heating limestone to temperatures over 1400 °C, which frequently requires use of fossil fuels. The process is called calcination. When limestone is calcined it breaks down into lime and carbon dioxide. The latter is released into the atmosphere where it traps heat and is a contributor to global climate change. Taken together, the carbon dioxide produced by calcination, added to that produced by the fossil fuels that are used for calcination, combined with the fossil fuels used in transportation amounts to a staggering 1.5 giga tons (Gt) of CO_2 each year. This is equivalent to the CO_2 emissions from over forty typical coal-fired power plants. As many countries around the world, most notably China and India, use increasingly larger amounts of concrete the quantity of CO_2 produced by this industry is likely to continue to grow.

Research at Washington State University has shown that the net amount of greenhouse gases produced by cement may actually be lower than previously thought if the entire lifecycle of the material is considered. Over time small quantities of CO_2 reabsorb into concrete, even decades after it has hardened [7]. Although the exact quantities of carbon dioxide that can be sequestered in cement and the processes by which this happens are not yet well understood, the results are encouraging, as there appears to be no slowing down in our use of this building material.

Among other approaches to reducing the carbon footprint of concrete, scientists and entrepreneurs are looking for "greener" cements. A process developed by California company and Silicon Valley start-up Calera would harness carbon dioxide emitted from a power plant and mix it with seawater to create carbonates that are then used as a replacement for limestone in the manufacture of cement. The process proposes two key advantages: capture of CO_2 that would otherwise enter the atmosphere—reuse of a waste product—and reduction in the amount of calcined lime needed [8].

Some concrete companies are looking at additives that can be used as fillers to "bulk up" the mixture thereby reducing the amount of cement that is required. Some of the additives that have been evaluated are slag, the glassy layer that forms during the production of steel and fly ash, a powdery residue from burning coal [9]. Fly ash is a complex chemical mixture consisting primarily of oxides of silicon, aluminum, iron. It is also a source of calcium, which can substitute for the energy-intensive calcined lime. Magnesium, potassium, sodium, titanium, and sulfur may also be present, which adds to the complexity. The exact composition depends on the type of coal that is burned.

The use of fly ash in concrete has been shown to provide both performance and environmental benefits. For example, adding fly ash can make it last longer. Fly ash containing concrete has been used in applications that are exposed to severe weather conditions such as the decks and piers of Tampa Bay's Sunshine Skyway Bridge in Florida. Completed in 1987, the Sunshine Skyway is the world's longest cable-stayed concrete bridge and has to routinely battle hurricanes, tropical storms, and sweltering temperatures. By replacing or displacing manufactured cement in concrete there is net reduction in energy use and greenhouse gas emissions together with a conservation of natural resources.

Concrete is the most widely used manufactured material on the planet. Its permanence made it the foundation of the ancient world and the demonstration of a country's place in the modern world. The Empire State Building in New York City, the Burj Khalifa in Dubai, and the CITIC Plaza in China all reflect a country's importance as an international economic powerhouse—a statement of grandeur.

Notes

1. This is quoted from The Repertory of Patent Inventions and other Discoveries and Improvements in Arts, Manufactures, and Agriculture, July—December (1825) p. 453. The specific wording in the patent is slightly different.
2. The Romans were fortunate in having available a volcanic earth called *pozzolana*, (after Pozzuoli, where it was first discovered. When mixed with lime (CaO) pozzolana formed a cement that resisted both fire and water.
3. The original seven wonders were: Colossus of Rhodes, Great Pyramid of Giza, Hanging Gardens of Babylon, Lighthouse of Alexandria, Mausoleum at Halicarnassus, Statue of Zeus at Olympia, and Temple of Artemis at Ephesus, of which the only one standing is the Great Pyramid.
4. ASTM C150 is the Standard Specification for Portland Cement.
5. Cited in https://www.smithsonianmag.com/science-nature/building-a-better-world-with-green-cement-81138/
   Accessed 20 May 2020.
6. Washington consumers have the happy distinction of paying the lowest price for electrical energy in the United States at only 7.4¢ per kWh. The national average for the country is 10.4¢ per kWh. Electric Data Browser (http://www.eia.gov/electricity /data/browser) Energy Information Administration Washington DC. Values from June 28, 2016.
7. A detailed letter Wilkinson wrote to *The Builder* that was published in January 1884 shows that he understood the importance of keeping the concrete in compression and using iron to support tensile stresses. A large part of the letter is reproduced in Newby, Frank (2016) *Early Reinforced Concrete, Studies in the History of Civil Engineering* Volume 11, Routledge.
8. Since 2008 the world's tallest structure is the Burj Khalifa in Dubai, United Arab Emirates with a total height of 829.8 m (2,722 ft). Construction of this megatall skyscraper used 431,000 cubic yards of concrete and 61,000 tons of "rebar." Rebar, which is short for reinforcing bar refers to the steel bar or mesh that is used to reinforce concrete. The special feature of rebar is the surface ridges (like Ransome's twisted iron rods) that improves adhesion between the metal and the concrete.

9. The raw materials in concrete are inexpensive. However, the process to produce cement is very energy intensive and has a large carbon footprint.
10. International Council Report on Monuments and Sites to the World Heritage Committee.

References

1. Mark, Robert and Hutchinson, Paul (1986). On the structure of the Roman Pantheon. *The Art Bulletin, 68*, 24–34. The authors conducted modelling of the Pantheon structure with and without cracks to determine the effect of tensile and compressive forces.
2. Jennings, H. M. (2000). A model for the microstructure of calcium silicate hydrate in cement paste. *Cement and Concrete Research, 30*, 101–116. This is Jennings's most well known and cited paper (917 citations as of February 23, 2021.) The model described became known as the "Jennings Model."
3. Jackson, M. D., Mulcahy, S. R., Chen, H., Li, Y., Li, Q., Cappelletti, P., & Wenk, H.-R. (2017). Phillipsite and Al-tobermorite mineral cements produced through low-temperature water-rock reactions in Roman marine concrete. *American Mineralogist, 102*, 1435–1450.
4. Weibull, Waloddi (1951). A statistical distribution function of wide applicability. *ASME Journal of Applied Mechanics*, September, 293–297. Although Weibull applied his model to the failure probability of materials it actually has a broad applicability. One additional example in Weibull's original paper was applying the model to the statures for adult males born in the British Isles.
5. Kurrer, Karl-Eugen (2008). *The History of the Theory of Structures: From Arch Analysis to Computational Mechanics*. Berlin: Ernst and Sohn.
6. Frampton, Kenneth & Futagawa, Yukio (1983). *Modern Architecture 1851–1945* (p. 116). New York: Rizzoli.
7. Haselbach, Liv (2009). Potential for carbon dioxide absorption in concrete. *Journal of Environmental Engineering, 135*(6).
8. http://www.calera.com/beneficial-reuse-of-co2/products.html Accessed February 2, 2019.
9. Pedro D., de Brito, J. & Evangelista, L. (2018). Durability performance of high-performance concrete made with recycled aggregates, fly ash and densified silica fume. *Cement and Concrete Composites, 93*, 63–74.

8

Rubber—*The Material of Possibilities*

The Teatro Amazonas—The Amazonas Opera House—has an imposing pink façade and a dome comprising thirty-six thousand pieces of enameled pottery and vitrified tiles, depicting the Brazilian flag (Fig. 8.1). Inside this spectacular building is glass—including 350 lamp shades—made by the best glassmakers in the world working in Murano, a small island less than one mile north of Venice. The highest quality steel was imported from Liverpool, England for the internal columns and the finest Italian marble was brought from Rome to line the walls of the opera house. Great European artists including Domenico DeAngelis and Giovani Capranesi are responsible for the beautiful paintings in the Ball Room and in the Auditorium. The stage curtain illustrated by Brazilian painter Crispim do Amaral depicts the local water goddess Iara amid the nearby "meeting of the waters" where the warm black water of the Rio Negro meets the cooler sandy colored Rio Solimoes.

I am sitting in one of the exclusive boxes on the third level of the house. My elbows are resting on the tired and worn velvet-covered handrail. The play is *Chuva Constante*, an intense story of two cops, childhood friends, whose relationship is strained by personal struggles and alcohol, but I can understand no more than a word or two, as it is in Portuguese, a language I don't know. One hundred and twenty years ago I would not even have been able to afford the price of a ticket in one of these once plush boxes. Instead, the person who would most likely be occupying my seat—and maybe entertaining a business partner or having a secret romantic liaison—would

Fig. 8.1 The iconic Teatro Amazonas in Manaus designed by Celestial Sacardim

have been a "rubber baron," a name given to the rich owners of the rubber plantations in Brazil.

Manaus, situated on the left bank of the black waters of the Rio Negro, is the capital of the state of Amazonas and is the gateway to the Brazilian Amazon. In the late nineteenth century, Manaus was transformed from a forgotten Amazonian backwater into an economic and cultural metropolis when it became the rubber capital of the world. Cargo ships that plied the waters of the mighty Amazon from Belem to Manaus and back were loaded with rubber for Europe and North America. On their return journeys, the ships would bring back the trappings of Western civilization, including the materials to construct the opulent and very beautiful Teatro Amazonas.

Between 1908 and 1910—the apogee of the rubber boom—80 million rubber trees (*Hevea brasiliensis*) were spread over one and a half million square miles surrounding the city. Each year these trees produced 80,000 tons of raw rubber. This was an enormous part of not just the region's economy, but of the economy of the entire country.

The sticky sap that slowly oozes from the *Hevea brasiliensis* tree when the bark is cut was a material well known to the indigenous people in Central and South America well before the arrival of European explorers. Heat is required to turn the milky emulsion, called "latex," into solid rubber. A simple way to produce rubber is to dip a wooden paddle into a bowl containing the liquid latex and then dry it over a log fire. Slowly the water is driven off leaving a rubber coating on the paddle. Repeated dipping and drying allows the rubber layer to thicken.

A rubber ball can be obtained by drying the latex on a rounded mold of mud or clay. By the thirteenth century articles of rubber, including balls for games, were in common use among the Mayas and Aztecs. Children in Haiti—just like children around the world today—used bouncy rubber balls to play games. The balls could be thrown, caught, and kicked. But not all ball games in the ancient world were so lighthearted. Pitz, a classical Mayan ball game was an extremely fierce encounter that, despite the protective equipment that was worn by the competitors, frequently involved serious injuries and even death if the player was hit by the heavy solid latex ball. But as described by Douglas Preston in *The Lost City of the Monkey God*: "The ball game was a vital Mesoamerican ritual, and playing it was essential to maintaining the cosmic order and keeping up the community's health and prosperity" [1].

In addition to balls, the indigenous people had found other uses for what was called "hevea" or its Maya name, "caoutchouc": it was used as a waterproof covering for footwear and clothing, as an adhesive, and to make watertight containers that could be used for holding liquids. Some of these applications for rubber were to be rediscovered by Europeans much later and become important commercial products for this stretchy and deformable material.

Columbus returned to Europe in 1496 with some caoutchouc, but no Europeans found any useful applications for it. It largely remained a curiosity if people even thought about it at all.

Over two hundred years later, in 1735, French naturalist and mathematician Charles-Marie de la Condamine was sent by the French Academy of Sciences on an expedition to Peru to make measures of longitude near the Equator. On this trip de la Condamine was interested in cinchona, the bark of which is the source of the antimalarial drug quinine. While there, de la Condamine observed the link between caoutchouc and the *Hevea brasiliensis* tree. In February 1745 the probably weary traveler returned to France and brought back samples of the material together with information about its properties, which was presented to the French Academy of Sciences and eventually published in 1755.

The Western history of rubber then took a hiatus until 1770, when English chemist Joseph Priestley observed that caoutchouc could erase or, as the English say, "rub out" the marks made by a black lead pencil. Priestley wrote in his book *A Familiar Introduction to the Theory and Practice of Perspective*: "I have seen a substance excellently adapted to the purpose of wiping from paper the marks of a black-lead-pencil. It must, therefore, be of singular use to those who practice drawing. It is sold by Mr. Nairne, Mathematical Instrument-Maker, opposite the Royal-Exchange. He sells a cubical piece, of about half an inch, for three shillings; and he says it will last several years" [2]. Priestley is generally credited with the naming of rubber because of his observation of its erasing properties of pencil marks on paper. We might wonder: if an American had made Priestley's discovery, would the material have been known as "eraser"? Natural rubber was also known as India rubber because of its origin in the West Indies.

Naming aside, for Europeans rubber still remained an interesting, but not particularly useful, material until the middle of the nineteenth century. In the early 1820s English inventor Thomas Hancock patented a number of ideas for how to use and process rubber. Hancock described his surprise that "a substance possessing such peculiar qualities should have remained so long neglected, and that the only use of it should be that of rubbing out pencil-marks" [3]. The most valuable and unique property that Hancock ascribed to rubber was its elasticity: a property that no other material possessed or could be substituted for. One of Hancock's earliest innovations, which he patented in April of 1820, was to combine rubber imported from South America with leather or cotton to form a springy material that could be used in several forms of clothing including suspenders, the wristbands of gloves, and slip-on shoes.

As his business grew, Hancock had difficulty procuring sufficient rubber of the right size to work with. Another problem he encountered was that the cutting process was very wasteful. To be more efficient Hancock needed to be able to work with whatever sizes of rubber were available and to reuse the offcuts, relying less on a continuous supply of fresh material.

A major challenge with obtaining and using it is that latex coagulates very quickly, so it can only be formed into useful shapes close to its source. Transporting latex to a processing site could cause it to become unworkable. What was needed was a solvent that dissolved rubber allowing it to return to its original liquid state. French chemists proposed that turpentine or ether could be used. Turpentine was particularly abundant as it was used as an internal medicine and as an alternative to whale oil in lamps. Unfortunately, when the solutions were dried—particularly when turpentine was the solvent—the resulting rubber was of a poor quality and not as easy to work with as the original latex solution.

To address these problems Hancock made his most important and significant contribution to the history of rubber. He invented a processing machine called a masticator. The masticator could combine rubber of every size and shape including scraps into a homogeneous solid mass, which could be formed into large cylinders. The cylinders were so large in fact that two men were needed to carry them, but they could easily be compressed in iron molds to obtain blocks of rubber in any desired shape. Hancock even used thin sheets obtained from his masticator to produce threads, which were used for elasticated fabrics and webbing.

Around the same time that Hancock was working with his masticator, Glasgow chemical manufacturer Charles Macintosh was also experimenting with ways to incorporate rubber into various fabrics. Macintosh found that rubber could be dissolved in coal-tar naphtha—a volatile hydrocarbon liquid produced by distilling coal tar—and used as a glue to hold two pieces of woolen cloth together. The resulting double-layer fabric was waterproof and used to make raincoats, which still bear the name "mackintosh" (with a "k"). The first factory making mackintoshes was opened in Manchester in northern England and began production in 1824. By using Hancock's masticator to supply the rubber, Macintosh was able to produce his mackintoshes more economically, and the two men became partners in the manufacture of waterproof garments. Thanks to the efforts of Hancock, Macintosh and others, rubber manufacturing in Britain was becoming an established industry. By 1830 imports of raw rubber were 23 tons and these rose to over 300 tons by 1840. While Macintosh was making mackintoshes, other British firms were making waterproof boots and shoes from rubber or rubber-coated fabric as well as a whole host of miscellaneous engineering and surgical products.

American companies were also active in the early rubber industry. However, they soon encountered the second challenge associated with the use of natural rubber: its properties vary greatly with changes in temperature. The raw material becomes very sticky and smelly in hot weather, and in colder climates or during harsh winters it is prone to becoming brittle and cracking. The discerning American consumer—maybe unlike their British counterpart—was not happy with boots and shoes that changed properties depending on the weather. Items could be stiff and uncomfortable in winter and then in summer turn soft and sticky.[1] The challenge of stabilizing rubber over a wide range of temperatures and conditions preoccupied Philadelphia hardware merchant Charles Goodyear for nearly ten years before he discovered in 1839 how to use a combination of sulfur, white lead, and—very importantly—heat to make stable rubber products.

Goodyear was born in 1800 in New Haven, Connecticut. His obsession with rubber and how to improve it consumed him and his family for almost a decade beginning in 1834 when a branch of the Roxbury India Rubber Company rejected his idea for a new valve for rubber life preservers. During this period Goodyear's health suffered, he spent multiple periods in debtor's prison (he referred to it jovially as his "hotel"), and he became dependent on relatives and friends for food and shelter. His eventual breakthrough was a classic case of pseudoserendipity.[2]

Using his wife's kitchen and assorted pots and pans Goodyear mixed a variety of chemicals into raw rubber. Nitric acid, lime, and turpentine were all added at one time or another, but none was able to make a satisfactorily stable rubber. Another chemical that was in Goodyear's makeshift laboratory was sulfur. Powdered sulfur was well known as a fungicide and pesticide so would have been a widely available household chemical at the time. During an experiment where he was mixing rubber and sulfur, which hadn't previously proved to be a particularly useful combination, the mixture accidently touched the surface of a nearby hot stove. To his surprise the rubber didn't melt, even though the stove was hot enough that the rubber should have been transformed into a soft sticky mess. Goodyear immediately realized the significance of what had just happened - he had changed the properties of the rubber. His daughter commented: "As I was passing in and out of the room, I casually observed the little piece of gum which he was holding near the fire, and I noticed that he was unusually animated by some discovery that he had made. He nailed the piece of gum outside the kitchen door in the intense cold. In the morning, he brought it in, holding it up exultantly. He had found it perfectly flexible, as it was when he put it out" [4]. Through a series of further tests Goodyear was able to optimize the temperature and time of the process: 270 °F for between four to six hours. He applied for a patent for a process he called *vulcanization*, after Vulcan, the Roman god of fire, which was granted in 1844 [5]. Hancock obtained a British patent for vulcanization in November 1843, a couple of months before Goodyear's application.

By providing a method to stabilize rubber, vulcanization enabled the manufacture of a wide variety of goods from flexible rubber hoses and gloves to combs, battery jars, flooring, and much more.

So, what exactly is rubber, and how does vulcanization stabilize it? The primary chemical constituent of natural rubber, the form extracted from the *Hevea brasiliensis* tree and its synthetic equivalent, isoprene rubber, is a polymer called polyisoprene. The structure of any polymer is complex because they consist of such enormous molecules. The isoprene molecule—the basic building block of polyisoprene—consists of five carbon and eight hydrogen atoms and has the molecular formula, or structural unit, C_5H_8. The carbon atoms are oriented in a zig-zag fashion. This configuration is typical in chains of carbon atoms and is the result of the electronic structure surrounding each carbon nucleus. Connecting many isoprene molecules together results in a very long chain molecule, polyisoprene. "Poly" simply means "many", so natural rubber consists of many isoprene molecules.

In a piece of natural rubber, the polyisoprene chains form a jumbled, disordered arrangement: think strands of cooked spaghetti in a bowl. When the rubber is stretched the chains begin to align and the arrangement becomes increasingly ordered: think uncooked spaghetti in a box. When we stop stretching, the structure of the rubber, both that which we can see with the naked eye and what happens at the microscopic level, returns to its original shape. This is the characteristic elasticity of rubber: you can stretch it, you can squash it, and it will bounce back.

During vulcanization, sulfur atoms are added to the polyisoprene structure. These atoms connect—cross link—adjacent polymer chains on average once for every 1,000 carbon atoms.[3] The elasticity, stretchiness, of vulcanized rubber is then determined by the number of crosslinks, which in turn is related to how closely the sulfur atoms are spaced along the polyisoprene chains [6]. The elasticity, stretchiness, of vulcanized rubber is then determined by the number of crosslinks, which in turn is related to how closely the sulfur atoms are spaced along the polyisoprene chains. Low sulfur concentrations leave the rubber soft and flexible. Increasing the amount of sulfur makes the rubber harder, more rigid, and eventually brittle. If sufficient sulfur is added we now have "hard rubber," or Ebonite, which was a product created by Charles Goodyear and his brother Nelson. One of the applications for hard rubber was uniform buttons during the mid-nineteenth century for the U.S. Navy and Army. All rubber buttons made using the Goodyear patent bear the Goodyear name and patent date as well as the name of the manufacturing company, such as The Novelty Rubber Co. (N.R. Co.). Hard rubber buttons from the Civil War period are highly collectible.

By the time Hancock's patent expired in 1858, vulcanized rubber was a commercial success. The value of the American rubber goods market was about $5 million. About $150 million in today's money. Products made from rubber such as raincoats, shoes, valves, belts, hoses, and tires were in ever growing demand. The benefits of rubber—particularly its elasticity—were much appreciated by passengers travelling by carriage and by train. Rubber springs and blocks made any journey by road or rail a little more comfortable and evened out the bumps and jolts. But it was the rubber tire that revolutionized transportation and would eventually consume more rubber than all its other uses combined.

Thomas Hancock, of masticator fame, produced the first road-vehicle tires in 1846. They were made of solid rubber attached to a metal hoop. This was an important starting point, but the key development was that of John Boyd Dunlop several decades later.

Dunlop, a British veterinarian, was motivated by the desire to make an improved tricycle for his ten-year-old son. This led to the idea of the pneumatic—air filled, rather than solid—rubber tire that used compressed air to further dampen road vibrations and create an even smoother ride for his little boy. The pneumatic tire was formed by a series of canvas layers, called plies, covered by a rubber tread. Dunlop patented his invention in 1888 and it was awarded British Patent Number 10,607.[4] Initially Dunlop was heavily involved in commercializing the new technology. He formed a company, The Dunlop Pneumatic Tyre Co., and acted as technical director. But Dunlop was a better veterinarian than he was a businessman and it fell to an energetic entrepreneur from Dublin, William Harvey du Cros, to lead the expansion of the company. Du Cros became executive director of the Dunlop Pneumatic Type Co. and held that position until 1896.

Two years after Dunlop's patent had been awarded the British Patent Office became aware of an earlier patent that had been filed by Robert William Thomson in 1845. Thomson's patent, British Patent 10,990, pre-dated that of Dunlop by over forty years and also described a pneumatic tire. But du Cros's commitment to the success of The Dunlop Pneumatic Tyre Co. was not to be doused. He saw the enormous possibilities of the pneumatic tire and bought up any competing patents including Thomson's.

Despite the disappointment that Dunlop must have felt at having his patent voided, his son surely appreciated his much-improved tricycle and the company Dunlop founded was to become a great commercial success. Dunlop's work led to the popularity of the bicycle and was to impact the automobile industry, which changed the world forever. Harvey Firestone, son of the founder of The Firestone Tire & Rubber Company, stated the obvious but nonetheless important connection between the car and the rubber industry: Without rubber there could be no tires, and without tires there could be no automobiles [7].

Many inventors have laid claim to the invention of the gas-driven car, but the credit is frequently given to German Karl Benz. In 1885 he made a motorized tricycle with a single cylinder gas engine, which had a "breath-taking" top speed of almost 10 mph. By 1893, Benz had moved up to a four-wheeled version. Both the three- and four-wheelers used solid rubber tires surrounding light, spoked wheels. Brothers André and Edouard Michelin first used pneumatic tires on a motor vehicle in June 1895 when competing in the Bordeaux-Paris motorcar trials. Unfortunately, the Michelin's in car number 46 finished in last place, but the tires apparently were impressive and launched the pneumatic automobile tire industry. Dunlop's version appeared five years later. Because of the extra weight and traction requirements, car tires require more canvas plies and a thicker rubber than bicycle tires, thereby creating an even greater demand for rubber.

When Henry Ford started mass-producing his Model T in 1908, owning a car moved within the reach of more people. By 1930, over 15 million Model Ts had been sold in the United States and each one of these had four pneumatic tires plus a spare: over 75 million tires! Today there are over 250 million cars on American roads and more than 1 billion cars worldwide. In Beijing alone, 1,500 new cars are added to the city's roads each day. So, it is not surprising that vehicle tires represent by far the largest market for rubber. About two-thirds of the 30 million tons of rubber consumed each year is just for tires.

First bicycles and then automobiles made rubber, in addition to a very useful material, a strategic commodity.

In 1900 Brazil, which was at the time the major supplier of the world's rubber, produced 27,650 tons (up from the 31 tons recorded in 1827). It was during this golden era for Brazil's rubber industry that construction of the Teatro Amazona began. The wealthy rubber barons dreamed of building a European city in the tropical climate in the middle of the Amazon rainforest. When the Teatro Amazonas was built, Manaus, now a sprawling city of two million people, was known as the "Paris of the Tropics." The European influence, which inspired the design of the Opera House, is rife throughout the architecture of the city.

Started in 1882 and opening on New Year's Eve in 1896, the Opera House, which now has seats for seven hundred became the cultural symbol of the city of Manaus. The Palace of Justice, the Adolpho Lisboa, and the Municipal Market, which was designed by Gustave Eiffel, all reflect the bittersweet relationship between the Europeans and the Manaós. The local guides talk proudly of their past as they point out the words "Francis Morton & Co. L^{TD} Engineers Liverpool" embossed on each of the steel pillars of the Municipal Market and the fact that Manaus was one of the first cities in Latin America to have electric streetcars. At the same time, they lament the decline in their city's prestigious economic and social position in the world after the rubber industry transitioned over 10,000 miles to the east.

Unfortunately, experience has shown us time and again that booms can turn into busts. That is exactly what happened to the rubber industry in Manaus. Rubber plantations in Malaysia, grown from seeds smuggled from Brazil thirty years earlier by Henry Wickham, an English planter living in the region, were more productive than those in the Amazon. It became increasingly difficult for the plantation owners in Manaus to find a market for their rubber and eventually they were forced to sell their assets at a loss to escape bankruptcy. The industry collapsed and the ships sailing to Europe were soon full not of rubber but people leaving the Amazon. Manaus went into a rapid decline, but the Teatro Amazonas—the centerpiece of the "Paris of the Tropics"—still stands proudly as a memory of Manaus's glittering past. In the Sao Sebastiao Plaza, in the shadow of the opera house, tourists and locals still gather late into the evening to drink caipirinhas and enjoy this incredible city in the middle of the Amazon.

As the demand for rubber increased with the growth in the production of the automobile there were many attempts to produce improved synthetic forms of this versatile material. By 1912, after half a dozen years of research, the Bayer Company in Germany was producing methyl rubber made by polymerizing the molecule methylisoprene. Methylisoprene has a very similar structure to that of isoprene, which is the molecule polymerized to make natural rubber. The production of methyl rubber was ramped up to a large-scale industrial level during World War I, when a blockade halted rubber imports to Germany. Many of the new combat technologies that were developed between 1914 and 1918, from airplanes to tanks, all relied on rubber, whether it was using the limited supplies of natural rubber or the recently produced synthetic materials.

The use of the car also accelerated during the war years. For instance, the iconic five-seater Vauxhall D-Type staff car, or "25hp" rolled off the south London production line of the Vauxhall Iron Works in 1915 was one of the defining vehicles of World War I. The D-Type crossed the battlefields of the Western Front (through Belgium, France, and Germany), Egypt, the Russian Empire, and Palestine. Its four-cylinder 3,969 cc engine generated 25 horsepower, which at the time was quite an impressive achievement. It could reach a top speed of just over 60 mph. By comparison Henry Ford's first Model T introduced in 1908 generated 20 horsepower and was capable of a top speed of 45 mph.

What was perhaps the most impressive characteristic of the D-Type, is that even with its skinny rubber tires it was able to deal with the appalling battle-scared road surfaces, which might have challenged even today's most rugged 4 × 4s. The D-Type offered an appealing and comfortable alternative to traveling by horse. At the beginning of World War I the horse was the go-to mode of transportation. Millions of horses were used to transport supplies, heavy weaponry, and troops. Every major army had a substantial cavalry. However, the tactics of trench warfare with its barbed wire and machine guns rendered cavalry attacks nearly useless, particularly on the Western Front. About 8 million horses died in the four years of fighting, and maybe it is safe to assume that the 1,500 D-Types—produced at the rate of one per day—used by the military kept the number of equine deaths from going even higher. Following the Armistice in 1918 the D-Type was the first vehicle to cross the Rhine. In December 1918 The *Auto Motor Journal* described the D-Type in the following, glowing terms: No tank is tougher; no battleplane is fleeter (relatively and in potentiality); no light cruiser is smarter to the eye …" [8]. Only two of these classic cars survive today. One of them (model number IC-0721) appeared in Steven Spielberg's 2011 film *War Horse*.

By the end of World War I, the car was replacing the horse not just on the battlefield but also as the preferred means of personal transportation. For instance, the last horse-drawn tram in New York City made its exit in 1917. With the growth in the number of cars, the rubber on which they relied for their tires became a major resource issue. Bayer's synthetic methyl rubber proved to be an expensive and inferior imitation of natural rubber and production was abandoned at the end of the war. But that was not the last of many attempts to develop synthetic rubber. The failure of methyl rubber and the introduction of export restrictions of natural rubber from British Malaya in 1922[5] really marked the beginning of major research and development programs around the world that eventually led to a range of polymers that would impact almost every possible aspect of our day-to-day lives.

In 1926 American chemists Joseph C. Patrick and Nathan Mnookin, seeking to develop an improved and inexpensive automobile antifreeze, discovered instead a process for synthesizing a synthetic rubber they named "Thiokol," from the Greek words for sulfur (*theion*) and rubber (*kolla*). This was the first synthetic rubber made in the United States. A company was established called the Thiokol Chemical Company and commercial production of this new material began in earnest in 1930. Thiokol shares many of the general properties of natural rubber such as its elasticity, even though it has a very different molecular structure. Rather than having long chains of carbon to carbon bonds, Thiokol has chains consisting of sulfur, carbon, and oxygen atoms. Such a structure is completely synthetic; it has never existed in the natural world. Thiokol was a product of both human ingenuity and a little luck.

As with many forms of rubber Thiokol demonstrated good resistance to chemicals, in particular oil, which made it a good sealant around automobile windscreens and as a lining in fuel tanks in the wings of airplanes. Scientists at the Jet Propulsion Laboratory in Pasadena, California discovered that Thiokol rubber made an ideal binder for solid rocket fuels. The rubber acted both as a binder and as part of the fuel—it burns. One example of an early solid rocket fuel was potassium perchlorate mixed with Thiokol rubber. This combination found use in harpoon guns, bazookas, and rockets. Despite a myriad of uses Thiokol rubber was not able to meet the holy grail for synthetic rubber. It was not a workable alternative to natural rubber for tires because it has poor heat resistance, a low tensile strength, and tears easily.

In 1930, the year commercial production of Thiokol began, scientists at DuPont's Experimental Research Station in Delaware using some published research performed by Father (and professor of organic chemistry) Julius Nieuwland of Notre Dame University, developed a synthetic rubber they christened "Neoprene." The molecular structure of Neoprene is very different from that of natural rubber. It is also very different from that of Thiokol because it contains carbon to carbon double bonds with a chlorine atom attached at one end.[6] But Neoprene does share the same oil resistance that was a characteristic property of Thiokol. Despite the high cost of Neoprene, DuPont quickly put it on the market where it found a number of industrial and domestic applications. The material went into full commercial production in 1934 after the Dayton Rubber Manufacturing Co. in Dayton OH, successfully tested it for tires.

While synthetic rubber programs were active during the interwar years in the United States similar developments were happening in Europe. German chemists at I.G. Farben, spurred on by Adolf Hitler, had developed a series of synthetic rubbers called Buna. These materials marked a very significant milestone in the creation of synthetic rubber because these were the first synthetic "copolymers". In a copolymer the long molecular chains that are characteristic of all polymers, consist of two or more types of monomer—rather than just a single type, as in polyisoprene—connected together. Copolymers are not only created within the laboratory they are also found in nature. One example is the extraordinarily strong dragline silk from orb weaving spiders, which is a copolymer of two proteins: a structure that has remained unchanged for more than 125 million years [9, 10]. An advantage of using copolymers, as nature figured out, is that it is possible by selecting the appropriate monomers to modify the properties of the polymer to meet specific needs. As one synthetic example, copolymers can be produced that contain both hard and soft regions. The result is a material that has the elasticity of conventional rubber while being suitable for typical plastic processing [11]. In Buna S rubber one of the polymers is *bu*tadiene, the other is styrene, and sodium (*na*trium in German) is the catalyst used for polymerization. Subsequently known as styrene-butadiene rubber (SBR)—confusingly, is sometimes called butadiene-styrene rubber (BSR) in the literature—this copolymer was to become the most widely used of all synthetic rubbers, representing about one-half of the total world production.

The two molecules that make up SBR butadiene and styrene are in the ratio 3 to 1. When SBR is synthesized both constituents are mixed in water. As the polymerization reaction proceeds the two types of molecule link together along the polymer chain in a random manner. Consequently, some chains may have more butadiene and some chains may have more styrene, but overall the ratio of 3 to 1 is preserved in the final material. The individual polymer chains are then linked together using the vulcanization process—exactly the same approach that was used by Charles Goodyear.

SBR is used in large quantities in automobile and truck tires, generally as a replacement for natural rubber. SBR is hardwearing and crack resistant. Like natural rubber, SBR reacts over time with oxygen and tiny amounts of ozone in the air. The big difference is that the reaction with SBR increases the interlinking of the chains, so unlike natural rubber SBR tends to harden with age rather than soften. The significant limitation of SBR is that it has low tear strength when hot. This is a problem in the most demanding of applications: aircraft tires.

In 1940, with World War II well under way in Europe, President Franklin Roosevelt realized that the United States was vulnerable because of its dependence on threatened supplies of natural rubber. He called rubber a "strategic and critical material" and created the Rubber Reserve Company (RRC), which was to stockpile reserves of natural rubber and regulate all synthetic rubber production. At the time the United States consumed about 600,000 tons of rubber annually. Only 0.4% of this was synthetic. The rest was imports of natural rubber. But there was an impending event on the horizon that was about to provide a big impetus for the United States to improve upon its earlier efforts and develop a low-cost, high-quality, general purpose rubber.

On December 7, 1941, the United States entered World War II. Only four days after Pearl Harbor, the use of rubber in any product that was not essential to the war effort was banned. The speed limit on U.S. highways was lowered to 35 miles per hour, in order to reduce wear and tear on tires. Americans were encouraged by the government to conserve rubber, contribute to scrap rubber drives, and abide by rationing regulations for tires, shoes, and other rubber products. Even President Roosevelt's pet Scottish Terrier Fala saw his rubber toys melted down and repurposed. This became the largest recycling campaign ever recorded.

Three months after the attack on Pearl Harbor, the Japanese invaded Malaysia and the Dutch East Indies desperate to take over natural rubber production from the Allies. This move gave Axis control to over 95% of world rubber supplies, plunging the United States into a crisis.[7]

Rubber was important for the war effort because:

- Over 1 lb of rubber is in every gasmask
- Each life raft required up to 100 lb of rubber
- A scout car needed over 300 lb of rubber
- One Sherman tank required 1,000 lb of rubber
- There was almost 2,000 lb of rubber in every heavy bomber

Furthermore, each warship contained 20,000 rubber parts and rubber was used to coat every centimeter of wire used in every factory, home, office and military facility throughout the United States [12].

A charge was sent to all chemists and engineers: The Allies need a synthetic alternative to meet the 600,000 tons, and growing, demand for rubber products. Much like the later demand for fissile material required by the Manhattan Project there were few detailed instructions on how the factories should prepare themselves to produce the needed materials. No facilities had been built and there were no established methods to produce enough of the necessary raw materials.

The challenge was enormous. The United States industrial sector had never been called upon before to shoulder such a massive task, to achieve so much so quickly. If the synthetic rubber program failed, the capacity of the United States, and its Allies, to fight and win the war would be dealt a major blow. Although in 1941 the outcome of this challenge was uncertain, the expansion of the American synthetic rubber industry that occurred was a remarkable achievement. The total production of Buna S rubber in 1941 was a little over 230 tons. By the spring of 1945 production of GR-S (Government rubber styrene—a similar recipe to Buna S and SBR) was 70,000 tons per month [13]! This was over six times more than the German output. The butadiene, which was an essential raw material for the production of GR-S was manufactured primarily from grain alcohol, which had an advantage over the coal-based acetylene that was used by the Germans. The American Midwest had abundant supplies of grain!

As might be expected, during the quest to develop a high-quality synthetic rubber there were a number of experiments that didn't quite produce the desired result. One of these approaches led not to a useful substitute for rubber, but to a material that became one of the most popular children's toys: Silly Putty®. A patent for Silly Putty, which was titled "Treating Dimethyl Silicone Polymer with Boric Acid" was filed at the United States Patent Office on March 30, 1943 by Rob Roy McGregor and Earl Warrick, both from Pennsylvania on behalf of Corning Glass Works in Corning, New York. As the title of the patent suggests McGregor and Warrick were trying to make synthetic rubber with silicone (the same material used in bathroom sealants). In one particular experiment, they added some boric acid to a gasoline-like fluid called Corning 200. After placing the mixture in an oven McGregor and Warrick left work for the day. The next morning when the oven was opened, they found inside was not a liquid but a strange solid substance, which could be molded, stretched, and was very bouncy. With the help of some clever packaging and marketing Silly Putty was sold in 28 g portions in a clear plastic egg-shaped holder. To date over 300 million eggs have been sold [14].

In most applications of a material the primary consideration is its properties. Steel cables are chosen to support suspension bridges because they are strong and tough. Aluminum is used in power transmission lines because it is light and an excellent conductor of electricity. Optical fibers are made of glass because it is transparent over the required range of frequencies. But in one of the unappealing applications of rubber its selection was, at least in part, political. The application was rubber bullets, which in the early 1970s were a symbol of The Troubles in Northern Ireland.

I have only ever once been on the receiving end of a rubber bullet. That was in 2014, when I was handed one by former Irish Republican Army (IRA) member turned peacemaker Jon McCourt. McCourt was visiting Washington State University to present a lecture at the Thomas S. Foley Institute for Public Policy. The rubber bullet was black. It was about an inch and a half in diameter and maybe four inches long. It looked like a large black candle with a smooth domed head. The 5 oz in my hand would feel a lot more if it came directly at you with a muzzle velocity of over 200 feet per second. My prior impression was that rubber bullets would be small and somewhat forgiving; just enough to be a deterrent, but not do any real damage. That was a long way from reality. Thomas Friel, Tobias Molloy, and 11-year old

Francis Rowntree were all killed in the early 1970s by rubber bullets fired by the security forces in Northern Ireland.

The British Army introduced rubber bullets (or baton rounds, as they are more generally known) in 1970 for riot control in Northern Ireland. Over the next five years the Army fired over 55,000 rounds.[8] The intended use of rubber bullets was that they would be fired at the legs of rioters or just in front of them so that the projectile would lose some of its momentum and bounce into the intended target. Often, they were fired directly at people at close range where they caused serious injuries and a number of deaths.

The original baton rounds developed in Singapore during the 1880s were short sections of wooden broom handles. Hong Kong police used baton rounds made of teak, which easily splintered and could cause fatal wounds. The rubber bullet was introduced not for "humanitarian reasons, or because of technical military considerations." It was introduced for political reasons. Wooden baton rounds, while considered acceptable by less liberal political regimes, were considered unacceptable for use within the United Kingdom. Sharp shards of wood sticking out of arms, legs, and heads was not what the government would want to see on the nightly television news.

The technical reason behind why rubber was used is very simple. It is the same as the reason that children in Haiti were using it for balls in the fifteenth century—it bounces! But the direction of the bounce is unpredictable, which means that when the bullet is fired there is no certainty as to where it will end up. It was clear during the short time that they were in service that rubber bullets were problematic, not least for those on the receiving end. The British government wanted an alternative. In 1975 the "less lethal" "plastic bullet" replaced the rubber bullet.

The replacement was a polymer called polyvinyl chloride, better known as PVC, the material used to make plastic pipes. The plastic bullet, which was lighter than the rubber bullet, was designed to be fired directly at a person from a greater distance. From a materials science perspective it is difficult to see what "advantage" PVC would have. Although the plastic bullet was lighter the higher muzzle velocity and correspondingly higher momentum would suggest that the extent of injuries if fired closer that the prescribed range could be even greater than that of a rubber bullet. Jonathan Rosenhead in an article for the *New Scientist* magazine in 1976 suggests again that the transition from rubber to plastic, like the transition from wood to rubber, was political. He proposes that "by 1976, the political and tactical nature of the war in Northern Ireland was changing. The climate was one in which the Peace Movement could take root and show some signs of sapping Catholic support for the IRA. Hence a more selective weapon, which could be aimed at the "ringleaders," could be less counter-productive" [15].

Rubber bullets are an example of a material that, like all materials, can be used for good as well as ill. Another application of rubber that also became linked with The Troubles is plastic explosive.

Plastic explosive—which should really be called "rubber" explosive—became widely used by engineers, an important weapon for the military, and a component in the arsenal of terrorists. A stick of dynamite is difficult to conceal. Plastic explosive, on the other hand, can be molded into any shape or concealed in any location. Layers of plastic explosive can be inconspicuously placed inside a brief case or backpack or molded to conform to the underside of a car. For drilling and mining applications plastic explosive can be packed into tight bore holes to allow controlled explosions.

The first plastic explosive was gelignite, invented in 1875 by Alfred Nobel: of Nobel Prize fame. Nobel had previously invented dynamite and made a fortune from the manufacture of armaments. Gelignite was very different from dynamite. Rather than incorporating the explosive nitroglycerin in an inert powder (Nobel used diatomaceous earth, which as mentioned in Chap. 1 is a great absorber of liquids), which was packed into tubes, in gelignite the nitroglycerine is combined with gun cotton, wood pulp, and saltpeter (sodium or potassium nitrate) to form a jelly-like substance. In this form it was easy—and safe—to mold.

As more powerful explosives were developed, there was a need to decrease the sensitivity so that the material could be safely handled and not explode accidently. There was also a demand, primarily from the United States military for formulations that were more flexible with a consistency resembling modelling clay that could be molded to control the direction of the resulting explosion. These objectives were achieved in an explosive known as Composition-4 (C-4), which went into production in 1960. It contains in addition to the explosive component RDX (an explosive nitroamine) polyisobutylene, a synthetic rubber.

Rubber went from a material with limited applications and apparently limited possibilities to be a vital part of transportation, aerospace, energy, electronics, and consumer products industries. The key to unlocking these possibilities was through the dogged persistence of Charles Goodyear. The properties of rubber are very sensitive to temperature, but vulcanization stabilized the structure and broadened the range of conditions where the material could be used.

The world market for rubber is 26 million tons with projections for continued growth. The major producers are Thailand and Indonesia. In

Jambi Province, Indonesia deforestation to increase rubber plantations has been criticized for destroying elephant habitat [16].

For SBR, the most widely used synthetic rubber, the market is currently is more than 2 million tons with a value of $8 billion. By 2025 the projections are that the market for synthetic rubber will reach $10 billion. The growing global demand from new automobiles as well as replacement tires is helping to drive the SBR market.

What to do with waste rubber is an issue that plagues all forms of polymer. Currently most of the world's waste rubber is incinerated. If it is done badly, as it is in some countries that receive waste rubber such as India, it can result in the release of carcinogenic toxins such as dioxins, furans, benzene, and heavy metals into the environment.

While the enabling innovation in the use of rubber was vulcanization, which involved insertion of sulfur molecules between the long molecular chains that make up rubber, there is now considerable interest in inverse vulcanization. This process involves insertion of organic molecules into sulfur [17]. An application for the resulting polymer is electrodes in lithium-sulfur batteries. This is an exciting possibility because lithium-sulfur batteries offer much higher gravimetric energy densities than the ubiquitous lithium-ion battery and they are cheaper.

Notes

1. The optimum temperature for rubber is 20 °C (68 °F). At temperatures of 5 or 6 °C (~40 °F) rubber becomes hard because it crystallizes. At high temperatures rubber will become sticky and not retain its shape.
2. In his book *Serendipity*, Royston M. Roberts, draws a clarification that Goodyear's accidental discovery of the vulcanization process of rubber is not serendipity, according to the strict definition of the term. Instead of finding something accidently that was not sought, he found a solution accidently that was desperately sought. Roberts calls such fortuitous accidents "pseudoserendipity." Roberts, Royston M. (1989). *Serendipity: Accidental Discoveries in Science*. New York: Wiley.
3. In Goodyear's sulfur-vulcanized rubber there are about 40–45 sulfur atoms per crosslink. In modern vulcanized rubber the number of sulfurs per crosslink can be as few as four.
4. Reports of Patent, Design, March 2, 1898 describes a patent infringement case between The Pneumatic Tyre Company, LD. and The Tubeless Pneumatic Tyre and Capon Heaton, LD. And Others. One of the cited patents was Dunlop's patent No. 10,607 of 1888. The case can be found at https://academic.oup.com/rpc/article-pdf/15/3/74/4606067/15-3-74.pdf Accessed 8 February 2019.

5. The Export of Rubber (Restriction) Enactment in October 1922 was introduced to stabilize rubber prices to protect British manufacturers. The restriction affected US companies as they were the major users of rubber.
6. Carbon to carbon double bonds are very strong. Almost twice as strong as regular (single) carbon to carbon bonds.
7. International Institute of Synthetic Rubber Producers, Inc., *Brief History and Introduction of Rubber*, IISRP, www.iisrp.com/WebPolymers/00Rubber_Intro.pdf cited in: Foo D.C.Y. (2015). The Malaysian chemicals industry: From commodities to manufacturing *CEP Magazine*, November pp. 48–52. The statistic is also cited in Srivastava, V.K, Maiti, M, Basak, G.C, & Jasra, R.V. (2014). Role of catalysis in sustainable production of synthetic elastomers *J. Chem. Soc. 126*, 415.
8. Sutton, Malcolm, *An Index of Deaths from the Conflict in Ireland 1969–1993*. Accessed through the CIAN (Conflict Archive on the Internet) web site at the University of Ulster. (http://cain.ulst.ac.uk/).

References

1. Preston, Douglas (1997). *The Lost City of the Monkey God* (p. 202). New York: Grand Central Publishing.
2. Priestley, Joseph (1770). *A Familiar Introduction to the Theory and Practice of Perspective*. London: J. Johnson and J. Payne, xv. Priestley also describes how cylinders can be drawn by starting with a parallelopiped and then wiping out certain lines.
3. Hancock, Thomas (1857). *Personal Narrative of the Origin and Progress of the Caoutchouc or India-Rubber Manufacture in England*. The book was first published in (1857). The version accessed is the digitally printed version published in 2014 by Cambridge: Cambridge University Press. The quote appears on p. 2.
4. Peirce, Bradford (1866). *Trials of an Inventor: Life and Discoveries of Charles Goodyear*. This book was originally published in 1866 shortly after the death of Charles Goodyear. It was republished in 2003 by the University Press of the Pacific. The quote is on p. 106 of the more recent edition.
5. Goodyear, Charles (1844). Improvement in India-Rubber Fabrics *US 3,633*. In the patent description Goodyear states that his invention will alter the properties of natural rubber such that it will not "become softened by the action of the solar ray or of artificial heat at a temperature below that to which it was submitted in its preparation … nor will it be injuriously affected by exposure to cold."

6. Flory, Paul J. (1944). Network structure and the elastic properties of vulcanized rubber. *Chemical Reviews, 35*, 51–75. Describes the molecular structure of sulfur-vulcanized rubber.

7. Firestone, H. J. (1932). *The Romance and Drama of the Rubber Industry*. Akron: Firestone Tire and Rubber Co.

8. *The Auto Motor Journal* (1918) Volume 23. December.

9. An, B., Jenkins, J. E., Sampath, S., Holland, G. P., Hinman, M., Yarger, J. L., & Lewis, R. (2012). Reproducing natural spider silks' copolymer behavior in synthetic silk mimics. *Biomacromolecules, 13*, 3938–3948.

10. Gatesy, J., Hayashi, C., Motriuk, D., Woods, J., & Lewis, R. (2001). Extreme diversity, conservation, and convergence of spider silk fibroin sequences. *Science, 291*, 2603–2605.

11. Feng, Hongbo, Lu Xinyi, Wang Weiyu, Kang Nam-Goo, & Mays Jimmy W. (2017). Block copolymers: Synthesis, self-assembly, and applications. *Polymers, 9*, 494. A review article on block copolymers. Gives a number of different applications that are possible by designing the materials to have specific properties.

12. United States War Production Board; U.S. G.P.O. America needs your Scrap Rubber (1942). Minnesota Historical Society. This 1942 poster shows the pounds of rubber needed to keep the military supplied.

13. Morris, P. J. T. (1994). Synthetic rubber: autarky and war in: *The Development of Plastics* (pp. 54–69). Cambridge: The Royal Society of Chemistry. The volume was edited by S.T.I. Mossman and P.J.T. Morris.

14. Thayer, Ann (2000). What's that stuff? Silly Putty. *Chemical & Engineering News, 78*, 27.

15. Rosenhead, Jonathan (1976). A new look at less 'less lethal' weapons. *New Scientist,* 16 December: 672–674.

16. Otten, Fenna, Hein, Jonas, Bonday, Hannah, & Faust, Heiko (2020). Deconstructing sustainable rubber production: contesting narratives in rural Sumatra. *Journal of Land Use Science, 15*, 306–326.

17. Simmonds, Adam G., Briebel, Jared J., Park, Jungjin, Kim, Kwi Ryong, Chung, Woo Jin, Oleshko, Vladimir P., Kim, Jenny, Kim, Eui Tae, Glass, Richard S., Soles, Christopher L., Sung, Yung-Eun, Char, Kookheon, & Pyun, Jeffrey (2014). Inverse vulcanization of elemental sulfur to prepare polymeric electrode materials for Li-S batteries. *ACS Macro Lett, 3*, 229–232.

9

Polyethylene—*The Material of Chance*

As Black Vultures circle overhead, Manaus's waste pickers (*catadores*) are sifting through mounds of garbage. While the vultures are using their keen eyesight to search for food, the waste pickers also are scanning through the piles of bottles, old clothes, cans, soiled cardboard and paper looking for anything of value that can be sold for recycling or reuse. Of the 20 million waste pickers worldwide, there are maybe half a million in Brazil—they are the country's recycling program [1]. Among the assorted debris of our disposable society is lots and lots of plastic. Plastic bottles and tubs that once contained shampoo, sunblock, Coca Cola, orange juice, yoghurt, or milk are mingled with flimsy plastic grocery bags and thin transparent films that covered meat, cheese, or fruit to stop it spoiling. The most abundant plastic in municipal solid waste dumps like those in Manaus is polyethylene—the most widely used plastic in the world. Annual production of polyethylene is now around one hundred million tons, far exceeding the amounts of aluminum and copper combined.[1] The world's top polyethylene producers include the familiar names of chemical and oil companies such as Dow Chemical and Exxon Mobil Corporation as well as Saudi Arabia's multinational chemical giant, SABIC and China's Sinopec Corporation.

The history of polyethylene begins like that of many polymers: an unplanned outcome of a scientific experiment.[2] In 1894 German chemist Hans von Pechmann discovered diazomethane, an explosive yellow gas that can be used in the synthesis of a number of important organic compounds. While experimenting with diazomethane von Pechmann found that it

© The Author(s), under exclusive license to Springer Nature
Switzerland AG 2021
M. G. Norton, *Ten Materials That Shaped Our World*,
https://doi.org/10.1007/978-3-030-75213-2_9

decomposed to form a white powder comprised only of carbon and hydrogen atoms. (Diazomethane contains carbon, hydrogen, and nitrogen.) This new material was named polymethylene but would later be called polyethylene to be consistent with naming conventions in chemistry.[3]

Almost forty years later, in 1930, two researchers at the University of Illinois, M.E.P. Friedrich and C.S. Marvel reported the results of a number of reactions involving organic arsenic compounds. At the conclusion of each experiment any residual compounds were described and where possible identified. During one particular experiment Friedrich and Marvel noted: "a white solid began to appear on the surface of the solution. This solid has not been studied further" [2]. That white solid was polyethylene, but it remained for a few years an unidentified curiosity.

The next part of the history of polyethylene takes us from the Midwestern United States to the north of England and Imperial Chemical Industries (ICI) in Northwich, Cheshire, lying almost equidistant between Liverpool and Manchester. In 1932 in the Alkali Division of ICI two chemists, Michael Perrin and J.C. Swallow proposed that the company study what happens when chemical reactions are carried out at high pressure. At this time, in the early 1930s, chemical kinetics was an evolving field. The goal of the proposed study was simply to gain a better basic understanding of how organic compounds—like polymers—formed. Over the next year many different reactions were tried and nothing particularly notable was observed. That was until March 1933 when Eric Fawcett and Reginald Gibson were studying the reaction between the molecules ethylene and benzaldehyde at a pressure of 1400 atmospheres (equivalent to the pressure at a depth of about 3 miles below the Earth's surface[4]). The reaction produced a "white waxy solid" on the walls of the reaction chamber.[5] Does the description of this product sound familiar to that reported by Friedrich and Marvel?

The polymerization reaction that led to the formation of polyethylene—polythene as it is often called in the United Kingdom—was due to oxygen impurities, the presence of which was unknown to Fawcett and Gibson at the time. Consequently, when they tried to reproduce the process with fresh reactants, they were unsuccessful. It took another two years before Michael Perrin was able to produce significant quantities of the new polymer. By 1939 ICI had begun production of a form of polyethylene called low-density polyethylene (LDPE) that became used around the world for plastic bags and plastic food packaging. LDPE can be identified for recycling purposes by the resin code identification number "4".

High-density polyethylene (HDPE)—the other common form of polyethylene—was discovered in 1951 by chemists Robert Banks and J.P.

Hagen working for Phillips Petroleum. One major advantage of HDPE from a manufacturing point of view was that it didn't require the very high pressures necessary to produce LDPE. HDPE is widely used to make containers that hold everything from milk to bleach and shampoo to dishwasher pods. It is also the most commonly used plastic for single-use grocery bags. HDPE is given number "2" in the resin code identification system.

Since its unintended synthesis in 1894 and eventual identification, polyethylene has become a critical and ubiquitous material in the modern world. It is essential insulation for the 200,000 miles of high-voltage transmission lines and the 5.5 million miles of local distribution lines, linking thousands of generating plants to factories, homes, and businesses [3]. Polyethylene film extends the shelf life of food helping to protect us from harmful bacteria and pathogens. A study on bananas found that after eighteen days unpackaged fruit was unmarketable, whereas bananas packed in polyethylene bags were good for over one month [4]. Most of the world's natural gas is distributed along polyethylene pipes. High-density polyethylene sheets provide a low permeable liner for sanitary landfill sites like the Apex Regional in Las Vegas, at 2200 acres the largest landfill in the world. The polyethylene limits toxic leachates entering and contaminating local aquifers and rivers.

The features of polyethylene that make it so useful and versatile are it is easy to process, it is inexpensive, and the raw materials are abundant (at least for the moment.) Furthermore, polyethylene is recyclable.

The majority of the early applications for polyethylene were as insulation for high frequency cables and, to a much smaller extent, in the manufacture of wax candles. With the onset of World War II, polyethylene, like rubber, became a material of strategic importance for the war effort. Consequently, use in non-essential applications, such as candles, was severely limited. (Candles were actually essential items in the trenches during World War I because they provided not only light, but also an effective way to kill lice.)

With the declaration of war with Japan in December 1941 the Post Office wanted to replace natural rubber with polyethylene as insulation for all telephone cables. The Japanese had control of the major rubber producing regions in Asia—which by this time had surpassed those in Brazil—and could easily block supplies of this critical material going to the Allies. An alternative material was urgently needed and polyethylene, if it could be manufactured in large enough quantities, might fit the bill.

The feasibility of this application had already been demonstrated by running a short telephone cable insulated using polyethylene between the picturesque Isle of Wight and the English mainland: a distance of about 10 miles. By the time the Japanese invaded Pearl Harbor production of polyethylene in the United Kingdom was happening around the clock—permission was even given for blackout restrictions to be ignored in these factories so that repair work on the equipment could happen at night allowing polymer production to resume first thing in the morning. The demand for polyethylene during the early years of the war increased sharply and by March 1942 more than 1000 lb a day could be produced [5].

In 1944, a number of polyethylene insulated telephone cables were laid across the English Channel to replace the original ones that had been cut early on in the war. The new cables provided heavy duty communication lines back to England from the large armies amassed on the Normandy beachheads to support the D-Day invasion, which foreshadowed the end of World War II, Hitler's dream of Nazi domination, and the Third Reich.

While telecom applications of polyethylene were of vital importance to the war effort, the bulk of polyethylene production went into the pivotal application of radar.

Radar is a technique that is used to identify objects by their ability to reflect radio waves. The basic principle of how radar works is quite simple. A transmitter sends out a series of radio waves (low energy electromagnetic radiation). When these radio waves hit an object, such as a plane or a ship, they are reflected back towards the receiver, which is typically part of the same device as the transmitter. Knowing the exact location and speed of an advancing enemy is a significant, and sometimes defining, tactical advantage in a battle. This is particularly true when bad weather limits visibility.

During the early development of radar, it had proved difficult to effectively insulate the equipment to prevent power loss and thereby preserve the strength of the returning signal. Polyethylene was the answer to the problem.

The excellent electrical insulation properties of polyethylene were well known from its use in underwater telephone cables. But polyethylene also has another important property. It is very light. With a density of only 0.9 g/cm^3 polyethylene is the lightest of all polymers. Nylon has a density greater than 1 g/cm^3, polyethylene terephthalate (PET) the plastic of soda bottles has a density of 1.4 g/cm^3, while Teflon has a density over 2 g/cm^3.

The low density of polyethylene insulation enabled the Royal Air Force to reduce the weight of radar to the point that the equipment could be placed *inside* their planes. This was one of the big secrets of World War II and provided an enormous technical advantage for the Allies in long-distance warfare, most significantly in the Battle of the Atlantic against

German submarines. Detection by radar-equipped aircraft could suppress U-boat activity over a wide area. Thankfully, the Germans were forced to use a bulkier insulating material for their radar, which was less effective.

By the spring of 1943, the British had developed effective sea-scanning radar, insulated with polyethylene and small enough to be carried in patrol aircraft armed with airborne depth charges. This technology was instrumental in sinking ninety, and damaging fifty-one, U-boats in the first year of its deployment alone [6]. Because the development of radar was so secret there is almost no information from this period about the quantities of polyethylene used for this technology.

However, after the war there was an excess of polyethylene and some of the manufacturers who had ramped up production to match the needs of the war effort were forced to close down. The question now was: how could this material that was so important for winning the war be used in peacetime? Several new household applications, such as polyethylene film lampshades were proposed, but none really took off until in 1947 ICI produced four-foot wide polyethylene films. One application for these thin, transparent, inexpensive, and tough films was immediately apparent—food packaging. It was made into bags for sliced bread and used as a covering to protect perishable deli products including meat and cheese. Polyethylene was approved by the Food and Drug Administration (FDA) as a "food contact material", because it was determined that components of the plastic would not migrate into the packaged food. This designation does not apply to recycled low-density polyethylene and some recycled high-density polyethylene products.

Agricultural products such as grain began to be transported in silo bags made out of polyethylene. White goods including refrigerators, washing machines and other appliances that filled the booming number of post-war homes were wrapped—as they still are—in polyethylene film to protect against accidental damage during transport. Polyethylene films are used in construction for house wrap. The film protects the house against moisture by not allowing water to get into the house while at the same time letting water vapor out to assist drying. The number of applications continued to grow until, not surprisingly films represented the single largest use of polyethylene.

Unlike rubber, the material featured in the previous chapter, there are no natural analogs of polyethylene. It is purely a synthetic material, created in a research laboratory and manufactured in a chemical plant. Also, unlike rubber, the history of plastics is relatively short, covering only about 150 years.

The story begins in 1863 in New York City. There was a worldwide shortage of ivory, the material of choice for billiard balls, because of the depletion of the wild elephant herds in Africa. The number of elephants had gone from 27 million at the turn of the century to 15 million in 1863 and would continue to decline over the next 150 years.[6] These majestic animals were the source of the precious ivory. The problem was further compounded by the increase in the popularity of billiards. A game that had once been an upper-class amusement over brandy and cigars was turning into a form of entertainment for the masses. As a result of these factors, New York company Phelan and Collendar, a major manufacturer of billiard balls, offered a prize of $10,000 for an ivory substitute that they could use to make billiard balls. It would also have the not inconsiderable benefit to the poor elephants by reducing the demand on the barbaric ivory trade. In 1863 $10,000 was a considerable monetary prize. Equivalent to about $200,000 today.

In neighboring New Jersey, a printer named John Wesley Hyatt and his brother Isaiah Smith Hyatt were experimenting with various materials in an attempt to create an ivory substitute. One of their experiments involved mixing sawdust and paper and holding it all together with glue. Fortuitously, as it turned out, during one of these experiments the tips of John Hyatt's fingers were getting raw and he went to get some collodion ("liquid cuticle" as it was commonly known at the time) to protect his skin. Collodion is a form of cellulose nitrate dissolved in ether and alcohol and was popular at the time, particularly with printers, as a type of do-it-yourself "Band Aid". Upon drying collodion forms a flexible colorless protective coating. To Hyatt's surprise, possibly annoyance, he found that his bottle of collodion had overturned, spilling the contents; the solvent had evaporated, leaving behind a hardened sheet of cellulose nitrate on the wooden shelf. Hyatt pulled of a piece of the dried film. He found it was tough and elastic and realized that this material might make a better binder for his sawdust and paper mixture than the glue he was currently using.

The initial results of using the collodion-containing material to make billiard balls were not successful. In fact, the results were literally explosive. A story printed in The Independent in 1917 describes the following incident highlighting the instability of Hyatt's formulation[7]: "A Colorado saloon-keeper wrote in to complain that one of the billiard players had touched a ball with a lighted cigar, which set it off and every man in the room had drawn his gun."

After some further experimentation, the Hyatt brothers found that a mixture of cellulose nitrate and camphor when heated under pressure (a shaping technique widely used in the modern ceramics industry and known as hot pressing), made a plastic suitable for billiard balls.[8] This plastic was named "celluloid" and patented in 1870. Although the new plastic was a great success and Hyatt formed the Celluloid Manufacturing Company in 1871, it seems he never received the $10,000 award from Phelan and Collendar for making an elephant-friendly billiard ball. The Smithsonian National Museum of American History in Washington D.C has an example of one of the Hyatt's celluloid billiard balls.[9] (Unfortunately, it is currently not on view to the public.)

The Independent article went on to say: "Here we have a new substance, the product of the creative genius of man, and therefore adaptable to his needs."

Indeed, celluloid became widely popular and was used for many applications that met the needs of both men and women. In the late nineteenth century, America was transitioning from a largely agricultural economy to one based on industry and manufacturing. Cities grew rapidly as workers moved from rural areas to jobs in new and expanding industries such as petroleum refining, steel manufacturing, and electrical power. Many of these workers needed smart, formal clothes, not the rugged jeans and boots used on the farm. Celluloid—a material that was transparent, flexible, and waterproof—provided an ingenious substitute for detachable linen or cotton collars, cuffs, and shirt fronts. Unlike their natural counterparts, the celluloid items never needed to be washed, starched, and ironed, they could simply be rinsed and worn again.

Although the concept of an easy-wear shirt using celluloid attachments sounded good there was always the possibility of a major wardrobe malfunction as the shirtfronts were prone to pop out unexpectedly: a lot of fun for the vaudeville comic, but not so impressive for the up and coming businessman. The stiff collar edges chafed cheeks and chins and celluloid buttons rattled noisily in the buttonholes of collars and cuffs. When shirt styles changed in the 1930s with the revival of the fixed collar shirt celluloid linens became a thing of the past.

Celluloid has the property associated with many plastics in that it can be formed into almost any shape and be made to look like almost any natural material. So, in addition to trying to save elephants, celluloid reduced the number of tortoises losing their shells to fashionable hair combs and trendy eyeglass frames. Artificial tortoiseshell made from celluloid was a very good and almost undetectable substitute. Celluloid could be made to look like the precious red coral of the Mediterranean, or colored to be indistinguishable from amber, or made cloudy to resemble onyx, or opalescent like alabaster. It could be molded into numerous objects from the luxury such as cutlery handles and fountain pens to the mundane of buttons and dice.

A polymer closely related to polyethylene and sharing a similar beginning as a chance discovery is Teflon, or to give it its full scientific name polytetrafluoroethylene (PTFE).

The building block of polyethylene is the ethylene molecule, C_2H_4. The carbon atoms form long chains that can contain many thousands of atoms. Each carbon is bonded to two other carbon atoms and to two hydrogen atoms. This structure is the simplest among all polymers. If the hydrogen atoms are replaced by fluorine, we get PTFE, which is built on the tetrafluoroethylene molecule, C_2F_4.

Teflon is a plastic that is familiar as the non-stick lining on frying pans and baking tins. Plumbers use Teflon, in the form of a slippery white tape, for sealing leaky joints in water pipes. But Teflon has many less visible uses as well. Surgeons use a porous form of Teflon called expanded PTFE for artificial veins and arteries.

The discovery of Teflon, like that of celluloid, polyethylene and a number of other polymers, owed a great deal to luck. Dr. Roy J. Plunkett a researcher at the DuPont research laboratories in Deepwater, New Jersey was working on chlorofluorocarbons (CFCs) to replace toxic chemicals like sulfur dioxide and ammonia as coolants in refrigerators and air conditioners.

These are the same CFCs that were once used ubiquitously as propellants in aerosol cans. The same CFCs that in 1974 were identified as the curse of the protective ozone layer that shields the Earth from the Sun's harsh ultraviolet radiation.[10] But in 1938 CFCs represented the possibility of a safer alternative in the lucrative refrigeration and propellant markets.

On the morning of April 6, 1938 Plunkett opened a cylinder that was supposed to contain tetrafluoroethylene and found that rather than holding 1000 g of gas it surprisingly only released 990 g. Where were the missing 10 g? On cutting open the cylinder he noticed that inside there was a white powdery residue, which he carefully scraped out using a metal wire and found that it accounted for the missing material.[11] What Plunkett had accidently created was a new polymer, which would become known and registered as Teflon® and be a multi-billion-dollar product for the DuPont company.

Even though DuPont was one of the leading polymer manufacturers in the world and had already filed patents for a synthetic rubber called Neoprene, a polyester fiber called Terylene, and Nylon, Plunkett's specific research project had not been to make a new polymer. But a revolutionary new polymer was exactly what he had made.

The superlative properties of Teflon soon became very evident and were reported for the first time in a paper published in the journal *Industrial and Engineering Chemistry* in 1941 by two other DuPont researchers [7].

In their paper Malcolm Renfrew and E.E. Lewis noted that "No substance has been found which will dissolve or even swell the polymer". They tested several hundred different reagents and found that "such potent reagents as aqua regia, … hot nitric acid, and boiling solutions of sodium hydroxide do not affect the polymer." Aqua regia is a powerful mixture of nitric acid and hydrochloric acid and notorious because of its ability to dissolve the precious metals gold and platinum. *The Spokesman-Review* the local paper from Renfrew's hometown of Spokane ran the headline "Plastic Defies Heat and Acids".

The reason for Teflon's incredible stability is that the carbon–fluorine bond is shorter than the normal C-F distance of 0.142 nm found in monofluorides. It is only 0.135 nm in polyfluorides. The difference is a miniscule, and seemingly insignificant 0.007 nm, but the extra closeness of the two atoms makes the bond rather formidable and very difficult to break.

The strength of the C-F bond is essential for giving Teflon its inertness and stability, but it makes fabricating it a challenge. In 1946 when Renfrew and Lewis wrote their classic paper on Teflon they lamented: "No general method has been found for bonding two pieces of the polymer together or for cementing pieces of polytetrafluoroethylene to metal." The latter problem was overcome by Marc Grégoire, an engineer at the French aerospace agency ONERA (Office National d'Etudes et de Recharches Aéronautiques.)

Grégoire was a keen fisherman and inventor of the fiberglass telescopic fishing pole. The challenge he had with his piscatorial pursuits was how to remove the composite material from its mold. From his work at ONERA Grégoire was familiar with the super slick properties Teflon and and decided it would make an excellent non-stick coating between pole and mold. The material worked so well that Marc's wife Colette suggested that he might apply a similar idea to create a non-stick pan by coating Teflon onto the cooking surface of an aluminum frying pan.

The challenge as Renfrew and Lewis noted was how to get a non-stick material to stick to something. Gregoire's approach was to roughen the metal surface creating anchor points where the polymer can attach. The adhesion is due not to a chemical interaction, but to mechanical interlocking in the same way that Velcro works; hooks on one side of the cloth catch in loops on the other side holding the two pieces of fabric together. There are several ways to roughen a metal surface. One approach is sandblasting, which involves bombarding the metal with sand particles that wear away tiny regions of the surface. Another is treating the aluminum surface with hydrochloric acid— the approach Grégoire used. The powerful acid etches tiny pits in the metal surface. After the surface has been etched the polymer is applied as an emulsion and heated to 400 °C for a few minutes. Upon heating the polymer melts and runs into the etch pits becoming wedged fast upon cooling to room temperature. Because the adhesion mechanism doesn't involve a chemical reaction between the Teflon and the aluminum over time the non-stick coating on a frying pan or baking tray begins to wear away and bare patches of aluminum become exposed.

Marc and Collette Grégoire formed a company Tefal (a contraction of the words *Tef*lon and *al*uminum) patenting the non-stick frying pan in 1954. The first Tefal factory was established in Sarcelles in the northern suburbs of Paris. During its first year in operation the factory produced frying pans at the rate of 100 per day.

Malcolm Renfrew later commented: "We knew it would be an important chemical, although it was not easy to fabricate. The frying pan thing … I would never have imagined that."

The inertness of Teflon, its inability to be dissolved even in the most corrosive acids gave it a critical role in the closing stages of World War II.

The discovery and successful synthesis of PTFE occurred just in time to play a decisive role in the development of the most powerful weapon ever created: a weapon that would hasten the end of the war and truly change the world forever. Roy Plunkett's invention contributed to the Manhattan Project, which was created in 1942 to produce the first atomic bombs that were then dropped on the Japanese cities of Hiroshima and Nagasaki in August 1945.

What was it about Teflon that made it so important in the production of the atomic bomb? It is resistant to the corrosive uranium hexafluoride, from which the fissionable isotope uranium-235 can be separated.

The production of uranium hexafluoride begins with uranium ore, the infamous "yellowcake". Through a series of chemical reactions, the ore is converted into uranium dioxide (UO_2), which is then reacted with hydrofluoric acid to produce uranium tetrafluoride (UF_4), and then again with fluorine to produce uranium hexafluoride (UF_6). Teflon provided the material for gaskets and reactor linings to handle the production of the corrosive materials and made the development of this devastating weapon possible. If Teflon had not been discovered it is possible that alternative routes to producing enriched uranium could have been found, but it would have been extremely challenging and possibly too late to obtain the 140 lb of uranium fuel that was used in Little Boy.

The debate on whether or not atomic weapons should have been used to end the war with Japan will probably never stop, but there seems little disagreement about whether or not we should have sent manned missions to the Moon. Teflon because of its extreme slipperiness and its stability over a very wide range of temperature, played a critical role in that milestone of human history.

Teflon is officially, according to The Guinness Book of World Records, the world's slickest substance. It has a coefficient of friction—the standard measure of slipperiness—of only 0.02: the lowest of any material. For comparison, the coefficient of friction of a waxed ski on snow is 0.05: twice as sticky as gliding on a Teflon surface. A car tire on asphalt has a coefficient of friction of 0.72: this, thankfully, is one of the highest measured values and is why we tend to stay on the road when driving. Teflon is so slippery that even geckos, which are well known for their ability to climb up vertical glass surfaces and to hang upside-down on ceilings, cannot hold on to a Teflon surface.

Lunar temperatures vary dramatically during a 12-h period. At night temperatures on the Moon can be as low as -183 °C (the temperature at which oxygen becomes a liquid). Temperatures soar to over 100 °C during the day (hot enough to boil water). Because of its unequalled stability over this enormous temperature range Teflon was the first material to set foot on the Moon. On 20 July 1969 at 10:56 pm Eastern Daylight Time when Neil Armstrong stepped from the lunar landing module onto the surface of the Moon, he did so wearing his Teflon boots. Although Armstrong and his colleague Buzz Aldrin did not experience the full temperature swing of the Moon—Armstrong recalled it being "really, really hot on the moon, 200 degrees Fahrenheit"[12]—their Teflon boots and space suits comprising an outer layer of Teflon-coated cloth would have been able to withstand the extraordinary range of temperatures. The astronauts, on the other hand, needed additional water-cooling! Teflon was also used on the knees, waist, and shoulders of the spacesuits. In addition to providing structural stability, the outer layer of Teflon protected the astronauts from micrometroids and lunar dust. These sharp particles can tear other fabrics, but when in contact with slippery Teflon they would simply slide off harmlessly.

Teflon a material that enabled the development of the atomic bomb, was the first material to set foot on the Moon and is found in almost every American household was the result of chance and a good dose of scientific curiosity. World production of Teflon has reached about 600,000 tons.[13] A very large number, but significantly less than the enormous quantities of polyethylene we use each year.

Celluloid, polyethylene, and Teflon all shared a healthy dose of serendipity in their discovery. Another characteristic shared by many plastics is that because they have no natural analogs they are not readily decomposed. The long chains of carbon–carbon bonds are resistant to degradation. Consequently, the amount of plastics that are part of our municipal solid waste facilities continues to grow. Of the 14 million tons of plastic waste discarded last year the largest component was LDPE and HDPE.

Once plastic was hailed as a wonder material—a material that could do anything. "The future is plastics" predicted Mr. Robinson in the movie *The Graduate*. Fifty years ago that dreamed of future was not of plastic pollution in every corner of the world. Littering beaches, floating in streams, strangling wildlife, inside our bodies, plastic is everywhere. Polyethylene, in particular, has gone from a material that helped to win the war and extend the life of perishable foods to being reviled, especially in its form of single-use plastic grocery bags. The Ministry of Ecology and Environment in China has issued a policy that would ban single-use plastic bags in all cities and towns by 2022. Other countries are considering similar steps.

Because polyethylene degrades very slowly in nature, to minimize its environmental impact recycling is essential. To encourage recycling and to make it easier for the different types of plastic to be correctly separated a classification was introduced by the Society of the Plastics Industry in 1988 that sorted common plastics into one of seven categories. The use of the now familiar triangle logo does not show what is actually recycled or even what can be recycled. Recycling rates of HDPE, which is usually assigned number 2 are only around 30% the highest of any of the seven categories.[14] Recycling rates of LDPE, identified by number 4, are significantly lower. Why are they not higher?

Recycling of plastics only works when the immediate economics are favorable. It is often cheaper to make new plastics even though the raw material is crude oil and the manufacture of ethylene (the precursor for polyethylene) is one of the highest CO_2 emitters among all industries. Many recycling processes involve "downcycling" where a lower grade product is produced. For instance, turning a HDPE shampoo bottle into a flower pot or LDPE food containers into plastic lumber. Each time the material is recycled it gets weaker because the length of the carbon–carbon chains decreases in a process called chain scission.

Our reliance on plastics is unlikely to decrease any time soon. These materials have become integral aspects of modern life. Current research suggests what the future might hold for polyethylene and other synthetic polymers. These include innovative approaches to degrading plastics, so they don't clog landfills and techniques for upcycling. An example of the former is a recent study looking at the possible degradation of polyethylene by caterpillars of the wax moth *Galleria mellonella* [8]. The researchers reported that 100 wax worms could, in a period of 12 h, decompose 92 mg of a LDPE grocery bag. (A typical bag weighs 5.5 g.)

Upcycling involves taking discarded polyethylene products and converting them into value-added liquid hydrocarbons, such as lubricants and waxes. In one recent study single-use polyethylene was converted into motor oil [9]. The process requires high pressure hydrogen, a temperature of 300 °C, and a complex catalyst consisting of platinum nanoparticles 2 nm in diameter supported on strontium titanate nanocuboids 50 nm in diameter. The results are exciting and major companies such as BP are looking at upcycling processes, but the economic feasibility has yet to be proven.

Notes

1. Worldwide aluminum production in 2019 was 64 million metric tons. Worldwide copper production in 2019 was 20 million metric tons. The latest annual production numbers are available online from the United States Geological Survey at usgs.gov.
2. We often use the terms "plastic" and "polymer" interchangeably. A polymer such as polyethylene is a long-chain molecule composed of repeating linked ethylene molecules. A polyethylene yoghurt tub for example is comprised mainly of the polymer, polyethylene, in addition to colorants (most yoghurt tubs are white) and stabilizers, which reduce the rate of degradation of the product.
3. The International Union of Pure and Applied Chemistry (IUPAC) has developed a standard naming convention for polymers. For simple polymers like polyethylene the name is derived from the source molecule. In the case of polyethylene that molecule is ethylene. So, polyethylene is "many ethylenes." The source molecule for polyethylene is not methylene, hence polymethylene does not satisfy IUPAC conventions.
4. Liou, J.G, Hacker, B.R, & Zhang, R.Y. (2000). Into the forbidden zone *Science 287*, 1215–16. This article from a group at Stanford University and University of California-Santa Barbara describes the pressure conditions below the Earth's surface.
5. Fawcett, E.W.G, Gibson, R.O, Perrin, M.W, Patton, J.G, & Williams, E. (1937). Improvements in or relating to the polymerization of ethylene *G.B. Patent, 471,590*. The British patent for polyethylene.
6. Data from https://www.futureforelephants.org/information/the-crisis-in-africa-elephant-numbers
7. Slosson, E.E. (1917). Plastics and elastics *The Independent*, December 22 p. 556. This was part of a series published in this newspaper under the title "Creative Chemistry: A Popular Explanation of Recent Progress in Chemical Industries".

8. http://americanhistory.si.edu/collections/search/object/nmah_2947   Accessed 12 February 2019. This website shows a picture of Hyatt's celluloid billiard ball. The billiard ball was a gift from the Celanese Corporation and is made of cellulose nitrate, a substance eventually known as celluloid.
9. http://invention.si.edu/imitation-ivory-and-power-play Accessed 12 February 2019. This is the website for the Lemelson Center for the Study of Invention and Innovation. It states that the Hyatt billiard ball with a celluloid base dominated the sport until the 1960s. The balls might not have been perfect but their use lasted for decades.
10. Molina M.J and Rowland, F.S. (1974). Stratospheric sink for chlorofluoromethanes: Chlorine atom-catalyzed destruction of ozone *Nature 249,* 810–812. F. Sherwood Rowland (1927–2012) and Mario J. Molina (1943–2020) shared the 1995 Nobel Prize in Chemistry with Paul J. Crutzen of the Max Plank Institute for Chemistry, Mainz, another pioneer in stratospheric ozone research.
11. This description of the discovery of Teflon is from John Emsley (1998) *Molecules at an Exhibition* Oxford: Oxford University Press, p. 132 and is supported by Plunkett's personal communication to Alfred B. Garrett (1962) The Flash of Genius, 2 Teflon: Roy J. Plunkett *Journal of Chemical Education 39,* 288.
12. Comments from Neil Armstrong to Robert Krulwich of National Public Radio. The link is http://www.npr.org/sections/krulwich/2010/12/08/131910930/neil-armstrong-talks-about-the-first-moon-walk
13. Global Polytetrafluoroethylene Market, 2014—2020, Zion Research Analysis 2015.
14. Data from the Environmental Protection Agency https://www.epa.gov/facts-and-figures-about-materials-waste-and-recycling/plastics-material-specific-data

References

1. Cass Talbott, Taylor (2019). A green army is ready to keep plastic waste out of the ocean. *Scientific American.* An online observation piece published October 7, 2019.
2. Friedrich, M. E. P and Marvel, C. S. (1930). The reaction between alkali metal alkyls and quaternary arsonium compounds. *Journal of the American Chemical Society, 52,* 376.
3. Weeks, Jennifer (2010). U.S. electrical grid undergoes massive transition to connect to renewables. *Scientific American,* online https://www.scientificamerican.com/article/what-is-the-smart-grid/.

4. Hailu M., Seyoum Workneh T., & Belew D. (2014). Effect of packaging materials on shelf life and quality of banana cultivars (Mus spp.). *Journal of Food Science and Technology, 51*, 2947–2963.
5. Wilson, G. D. (2006). Polyethylene: The early years in: *The Development of Plastics* (p. 81) edited by S.T.I. Mossman and P.J.T. Morris. London: Royal Society of Chemistry.
6. Hendrie, Andrew W. A. (2006). *The Cinderella Service: RAF Coastal Command 1939–194* (p.117). Barnsley, South Yorkshire: Pen & Sword Aviation.
7. Renfrew, M. M., & Lewis, E. E. (1946). Polytetrafluoroethylene: heat-resistant, chemically inert plastic. *Industrial and Engineering Chemistry, 38*, 870–877.
8. Bombelli, Paolo, Howe, Christopher J., & Bertocchini, F. (2017). Polyethylene bio-degradation by caterpillars of the wax moth *Galleria mellonella Current Biology, 27*, R292–R293.
9. Celik, Gokhan, Kennedy, Robert M., Hackler, Ryan A., Ferrandon, Magali, Tennakoon, Akalanka, Patnaik, Smita, LaPointe, Anne M., Ammal, Salai C., Heyden, Andreas, Perrasm Frédéric A., Pruski, Marek, Scott, Susannah L., Poeppelmeier, Kenneth R., Sadow, Aaron D., & Delferro, Massimiliano (2019). Upcycling single-use polyethylene into high-quality liquid products. *ACS Central Science, 5*, 1795–1803.

10

Aluminum—*The Material of Flight*

One hundred years ago, just less than two decades after the Wright Brothers historic flight, the National Advisory Committee for Aeronautics, the U.S. government's principal aeronautical research agency stated: " ... in the future all large airplanes must necessarily be constructed of metal".[1]

The metal is aluminum.

In current commercial airplanes, such as the Boeing 747 and the Airbus A320, like the Comet and Boeing 707 before them, aluminum accounts for much of their total weight. For instance, the Boeing 747–400, the latest in-service version of the instantly recognizable and iconic "Jumbo Jet" contains almost 150,000 lb (about 70,000 kg) of high-strength aluminum alloys: 80% of the weight of the empty plane. Aluminum alloys are everywhere; in the fuselage, the wing panels, the rudder, the exhaust pipes, the doors and floor, the seats, the engine turbines, and the cockpit instrumentation.[2]

In September 1968, when Boeing unveiled the 747 it was the largest passenger plane ever built. Measuring 232 feet (70.6 m) nose to tail with a wingspan of 195 feet (59 m) it could carry up to 550 passengers; more people than had flown in any plane before. And when the 747 in Pan Am livery flew across the Atlantic in January 1970 it became an icon of long-haul travel and exotic holidays.

The Concorde, the fastest passenger plane of all time, flying its manifest at over twice the speed of sound for 27 years, was built with an aluminum alloy skin, aluminum alloy wings, cockpit, engine intakes, and in the engine bay. The extensive use of aluminum allowed the Concorde to fly at more than

M. G. Norton, *Ten Materials That Shaped Our World*, https://doi.org/10.1007/978-3-030-75213-2_10

Mach 2.02 (approximately 1,350 mph) to the edge of space at a cruising altitude of 55,000 feet. On the ground, Concorde was 203 feet 9 inches long but stretched by almost 10 inches in flight due to heating of the airframe. The famous swing-nose reached a temperature of 127 °C—a stark contrast to the outside temperature of a subsonic aircraft of -50 °C. This high skin temperature rather than damaging the metal actually accounted for the excellent condition of the plane. Corrosive effects due to water vapor in the air were significantly reduced because the moisture would boil off the hot metal.

Aluminum was responsible for ushering in an era of low-cost, high speed, comfortable, even glamorous air transportation. In turn, we became a society obsessed with flying. The world has shrunk as countries as geographically separated as Iceland, New Zealand, and South Africa have become connected in a matter of hours and each has become a popular tourist destination because of long-haul air travel made possible by aluminum —more specifically, a range of aluminum alloys.

The first tiny pellet—the size of a pinhead—of aluminum was made in 1825 by the Danish scientist Hans Christian Ørsted. Ørsted extracted the aluminum by heating a mixture of potassium amalgam (itself a mixture of mercury and potassium) and aluminum chloride. The method was modified by the German Friedrich Wilhelm Wöhler in 1827 and then in the 1850s by the Frenchman Henri Sainte Claire Deville, who replaced potassium amalgam with its less expensive sodium equivalent. Aluminum now became available in marble-sized lumps, and by 1854 its cost had plummeted from $545 to $17 per pound, about $200 in today's money. Nowadays a pound of aluminum sells for about $1.

When aluminum was first produced it was available in such small quantities and cost so much that it was considered a precious metal alongside gold, silver, and platinum. Consequently, the earliest applications were where this exotic and rare new material could be proudly displayed. For instance, the apex of the Washington Monument, which was set on December 8, 1884, is made of aluminum. It weighs one hundred ounces when the cost of aluminum was about the same as that of silver and one ounce was equivalent to the daily wage of an average worker. Its value made it an appropriate finishing touch to the great structure, which was built to celebrate the legacy of George Washington, the first President of the United States. Upon completion the Washington Monument was the tallest building in the world, standing a little over 555 feet. And at the very top, no material was more highly placed, than aluminum. On the east facing side of the aluminum cap are the Latin words "Laus Deo," "Praise be to God."

On the other side of the Atlantic, the nude statue of "Eros" in Piccadilly Circus in London was cast in aluminum. Sir Alfred Gilbert sculpted the figure in 1893 to honor the 7th Earl of Shaftesbury who introduced the "Ten Hours Act" in 1833 limiting child labor and enforcing school education for children under thirteen. The statue is a London icon and a perfect meeting place for an evening at the theatre or in one of Chinatown's restaurants.

The principal problem with Deville's method was that it was expensive and inefficient in its use of raw materials: three pounds of sodium were needed to make a single pound of aluminum. What was needed was a cheaper method.

Both Sir Humphry Davy and Henri Sainte-Claire Deville had experimented using electricity to decompose aluminum oxide ("alumina") but the electrical currents needed were just too great to make the process cost effective. The breakthrough came in 1886 and happened simultaneously on both sides of the Atlantic. Working independently, Charles Martin Hall of Oberlin, Ohio and Paul Louis Toussaint Héroult of Paris, France developed a successful electrolytic cell for the production of aluminum from alumina. Hall filed his United States patent on July 9, 1886, while Héroult's French patent was filed almost three months earlier on April 23, 1886. Héroult also applied for a United States patent, but Hall was able to establish priority for the invention. Hall and Héroult showed not just a common interest in aluminum, they were both born in 1863 and both died in 1914. Although they were competitors, they became friends and the process used to manufacture aluminum became universally known by both their names.

The key to the Hall-Héroult process was finding a suitable solvent that would dissolve alumina. That solvent was cryolite, or sodium aluminum fluoride (Na_3AlF_6). Cryolite allows alumina to become molten at the relatively low temperature of 950 °C. It is impractical to try to melt alumina directly because its melting temperature is over 2,000 °C. Cryolite melts at 1012 °C and is an excellent solvent for alumina powder. The scientific principle in use here is called "depression of the melting point." The melting temperature of the cryolite-alumina mixture is lower than the melting temperature of either of the two components. It is for exactly the same reason that in winter we put salt on the sidewalk. The salt mixes with the ice and lowers its melting temperature. The ice melts below freezing (0 °C or 32 °F). It is also why the Chinese were able to make cast iron by melting a mixture of iron and carbon, which melts at a lower temperature than pure iron.

A Hall-Héroult cell is an example of an electrolytic cell using electrical energy to drive a chemical reaction. This process happening inside the cell is known as electrolysis. A Hall-Héroult cell, or "pot," consists of an insulated steel shell. The pots are lined with insulating blocks called fire brick followed by a thick layer of carbon blocks or casings. Carbon is a very good conductor of electricity and acts as one of the electrodes in the cell (much like one of the terminals in a battery). The carbon lining holds a pool of molten cryolite and alumina. Carbon rods, which form the other electrode or terminal, are dipped into the molten solution. A typical cell will have up to 20 carbon rods, each over one meter in length, arranged in two parallel rows.

An electric current flows down the carbon rods, through the molten cryolite-alumina mix and out through the cathode. The alumina is electrolyzed into aluminum and oxygen. For a typical cell the electric current might exceed 200,000 amps. The molten aluminum sinks to the bottom of the pot and the oxygen reacts with the carbon rods to form a mixture of carbon monoxide and carbon dioxide. Typically, every other day the aluminum is siphoned out of the cells. Each pot can produce about 500 lb of aluminum over a 24-h period. The carbon rods are slowly consumed as the electrolysis reaction proceeds requiring them to be replaced every ten days or so.

Before aluminum can be obtained from the Hall-Héroult cells it is necessary to first purify the mineral source, bauxite, and to convert it into alumina. A method for doing this was developed and patented by Austrian chemist Karl Joseph Bayer in 1888. In the Bayer process the raw bauxite is crushed and dissolved in sodium hydroxide (commonly known as caustic soda). The impurities in the bauxite that do not dissolve in the caustic soda are removed by passing the liquid through a filter. The filtered liquid is cooled, which causes pure alumina to precipitate as a fluffy white powder. Most of the world's aluminum is produced from Bayer process alumina using Hall-Héroult cells.

On September 18, 1888, using the technology patented by Hall, The Pittsburgh Reduction Company with $20,000 in private investments began producing aluminum at a rate of 50 lb per day at a set selling price of $5 per pound. Two years later the annual output of the plant had increased to 475 lb. In January 1907, the Pittsburgh Reduction Company became the Aluminum Company of America, then in 1998 Alcoa, the largest aluminum producer in the United States.

Lightweight materials are important in many applications. None more so than in all forms of transportation. In addition to revolutionizing air travel, aluminum played a critical role in the modernization of the railway industry. The construction of early locomotives involved primarily a combination of cast iron and wood. In particular the steam boilers were made from iron. Not only is iron heavier than aluminum it rusts. The boilers would often suffer from severe rusting because of the amount of boiling water that was necessary to power the trains. When the corrosion got too bad the boiler could explode.

The earliest use of aluminum in a train is the lightweight car built by the New York, New Haven, and Hartford Railroad ("The New Haven") in 1894 that used aluminum seat frames. In the 1930s, railroad companies including the Union Pacific and the Santa Fe experimented with all-aluminum railway cars, for both freight and passenger use. The Chicago World's Fair in 1933 featured the world's first main line aluminum passenger car the "George M. Pullman". This was named after the American industrialist and inventor, George Mortimer Pullman, who died in 1897 but whose name remains synonymous with luxury train travel.[3] The Pullman Company was proud to promote the amazing weight saving over steel cars: "A steel sleeper weights 180,000 lb. The George M. Pullman, with air-conditioning equipment, extra large generator, heavy batteries, and its water supply, tips the scales at 96,980 lb. Slightly more than half the weight, but with strength equivalent to all-steel construction! With no sacrifice of safety there is therefore an impressive reduction in the cost of haulage."[4]

In 1935, the *Comet* with an all aluminum alloy car body was commissioned by the New Haven Railroad. The *Comet*, which shares its name with the world's first jet airliner also made of aluminum, was built by Goodyear Zeppelin Corporation. The public had lost interest in airship dirigibles after a number of headline-grabbing crashes and the company was looking for new work for its aluminum airship factory. Making train cars were one such application.

By the 1960s aluminum was widespread in the railroad industry. And by the 1980s, aluminum was the material of choice for urban metro systems such as the Washington DC metro, which uses extruded aluminum alloys for the car body. The weight reduction in using aluminum compared to other heavier metals reduces energy costs and the lower inertia enables better acceleration between stops speeding up journey times. Combined with its excellent corrosion resistance aluminum offers low maintenance costs and a long system life [1].

Aluminum alloys are essential for high-speed trains such as Japan's Shinkansen (bullet train), which can reach operating speeds up to 320 km/h (200 mph) made possible by its lightweight construction. The Maglev Transrapid system, using magnetic levitation, developed in Germany is designed for speeds as high as 500 km/h. Lightweight materials make these high speeds achievable, but high construction costs have severely limited the extent to which maglev technology is used. The only commercially operating maglev train runs between Shanghai and Pudong International Airport, a distance of only about 20 miles. Briefly the trains reach a speed of 432 km/h (268 mph).

While for flight lightweight materials are critical for enabling planes to leave the ground, for railways the critical factor is lowering the amount of friction between the wheels and the track by reducing the weight of the rolling stock.

The many applications for aluminum are largely because of all the known metals it is one of the lightest. What makes aluminum so lightweight? Aluminum is element number thirteen in the Periodic Table of Elements. So, each aluminum atom contains only thirteen protons. These thirteen protons are usually combined with 14 neutrons, which give aluminum an average atomic weight of 27. Only the metals lithium, beryllium, sodium, and magnesium have fewer protons in their nuclei. However, even though these four metals are lighter than aluminum there are very good reasons why we don't use them in structural applications. Lithium—while essential for lithium-ion batteries and an occasional alloying element with aluminum—is very expensive and highly reactive. Beryllium is toxic; berylliosis is a fatal lung disease caused by inhaling beryllium dust. Sodium is soft and reacts violently with water making it difficult to handle. A common high school science demonstration is to place a tiny piece of sodium on a beaker of water and watch it fizz angrily across the surface! Magnesium—also an alloying element for aluminum—is expensive and the processes to extract it whether from seawater or its ore is very energy intensive.

Unlike all its lightweight competitors, aluminum is resistant to corrosion—a very useful property for both statues and airplanes—because it forms a very thin protective layer of oxide on its surface. This layer, which is much too thin to see with the naked eye, forms almost immediately upon exposing the metal to air. The so-called "native oxide" layer is just a few millionths of a millimeter thick, but it is enough to passivate the surface and prevent excessive corrosion. Immersing aluminum metal in sulfuric acid allows thicker oxide layers to be built up by a process known as anodizing. The oxide layers formed by anodizing can be up to 0.2 mm thick and are very hard (alumina is almost as hard as diamond—it is just one number lower on the Mohs hardness scale). Anodized aluminum finds many applications because of its hardness and because it can be dyed in almost every conceivable color from mocha tan to heron blue to the rose gold used in iPhones and MacBooks.

Aluminum is also surprisingly easy to mass-produce in thin complex shapes. The pop-tab soda can is just one example of a mass-produced aluminum product. Each year about 100 billion aluminum cans are produced in the United States alone. Over 30 cans can be made from a single pound of aluminum. Each can weighs just 14 g and has a thickness of only one quarter of a millimeter (about four times thicker than a human hair). The reason that aluminum can be formed so easily into complex shapes—such as pop-tab cans—is because of its highly symmetrical crystal structure.

Aluminum is the lightest element that exists in what is called the face-centered cubic (fcc) structure. This is one of the two close-packed arrangements where the maximum possible amount of space is filled with atoms. The other close-packed structure is hexagonal close packing (hcp). In both close-packed arrangements each atom is surrounded by twelve equivalent atoms. Each atom has twelve neighbors. There is no more efficient way of packing atoms (or any spherical objects including cannon balls and oranges) according to what is known as the Kepler conjecture [2]. The majority of metals are close-packed, including copper, silver, and gold.[5]

To describe what happens when a piece of metal, such as a metal sheet or wire, is bent to a new permanent shape—a process called plastic deformation—it is necessary to refer to a defect that is present in all metals including iron, described in Chap. 4. That defect is the dislocation.

Dislocations are line defects. They are rows of displaced atoms. During plastic deformation, as a force is applied to the metal, dislocations move from one location to another. When the force is removed it is impossible for the dislocations to move back *precisely* to their original position. Consequently, the metal has been permanently deformed. In close packed structures because the atoms are as close together as they can be the dislocations have shorter distances to move, which makes their movement easier. In other words, less force is required to move them. It can be compared to the difference between stepping over a stick to stepping over a fallen tree trunk. The tree trunk involves moving a greater distance and overcoming a larger barrier. The other important factor to consider in the case of deforming metals that consist of many grains is that in highly symmetrical structures it is generally easier for dislocations to move from one grain to the next, in other words the deformation can move from one grain to its neighbor (but not back again because the deformation is permanent).

To increase the strength of aluminum, or any metal, it is necessary to limit the movement of dislocations[6]. This is most often done by alloying. Alfred Wilm, a German metallurgist and engineer created an aluminum alloy called duralumin in 1903.[7] The main alloying elements in duralumin are copper (at slightly more than 3%) and 1% each of the magnesium and manganese. The excellent mechanical properties of duralumin are the result both of its composition—the type and amount of the alloying elements that are present—and the fact that it is heat treated during manufacture in a process called aging.

Aging considerably strengthens the resulting alloy by forcing the alloying atoms into the aluminum lattice at much higher amounts than are possible at room temperature.

The alloying elements cause a local disruption to the lattice because they are a different size (they are frequently larger) than the aluminum atoms. Because of the disruption of the crystal structure it is more difficult—requires more energy—to move dislocations through the alloy making it more difficult to deform and hence stronger. Over time some of the trapped copper atoms come out of the lattice and combine with aluminum atoms to form tiny crystals of copper aluminide ($CuAl_2$). These crystals cause even greater disruption of the aluminum lattice, which leads to a further enhancement of the strengthening effect.

So, is aluminum a perfect metal? It certainly has many very useful properties. It is easy to form into thin-walled complex shapes. When alloyed with other metals such as copper and manganese it is very strong: pound for pound, aluminum alloys are much stronger than steel. The major drawback of aluminum is that it does not exist in its pure metallic form in nature. Even though it is the most abundant of all metals (accounting for about 8% of the Earth's crust), it bonds so tightly with oxygen that most of the world's aluminum is tied up in the form of impure aluminum oxides, mainly bauxite.

However, for some applications, aluminum may be as close to perfection as it is possible to get with the choices of metals available to us. Aluminum, unlike its major competitor steel, has the ability even after more than a century to convey a sense of modernity, for instance the rose gold iPhone casing. The great architect and one of the founders of modernist architecture Ludwig Mies van der Rohe once asked, referring to aluminum, "Is there anything this material cannot do?".

Yet when aluminum was first produced it was actually difficult to find uses for it. Sainte-Claire Deville, one of the pioneers of aluminum synthesis, concluded in his book *De l'Aluminum*: "I have tried to show that aluminum may become a useful metal by studying with care its physical and chemical properties. As to the place it may occupy in our daily life that will depend on the public's estimation of it and its commercial price. The introduction of a new metal into the usages of man's life is an operation of extreme difficulty" [3].

As the material of flight, the use of aluminum in airplanes dates back to the very beginning of human flight and the pioneering work of Wilbur and Orville Wright. The Wright Flyer, which flew on December 17 1903 over the outer bank sands at Kitty Hawk, North Carolina, had a four-cylinder, 12 horsepower automobile engine, which the brothers had modified with a

30 lb aluminum block.[8] The use of aluminum rather than the heavier cast iron significantly reduced the overall weight of the plane allowing it to get airborne and fly with its single pilot. But wood—Sitka spruce and bamboo in the case of the Wright Flyer—was the mainstay of aviation materials into the 1930s.

During World War I, lightweight aluminum alloys began to challenge wood as the essential component for aerospace manufacture. Of the 170,000 military airplanes produced during that period the vast majority still used wood frameworks. In 1915 the German aircraft designer Hugo Junkers built the world's first full metal aircraft; the Junkers J 1 monoplane. Its fuselage was made from Wilm's duralumin alloy. Duralumin was of great advantage to Germany in World War I. It was very much stronger and less ductile than other aluminum alloys at the time and allowed the construction of superior aircraft and dirigibles. But airpower alone was, fortunately, not sufficient to win the war.

The interwar years between World War I and World War II came to be known as the "Golden Age of Flight." During the 1920s, American and European pilots competed in airplane racing with daring record-setting flights making national news first on radio then on television. This competitive environment led to innovations in design that produced ever faster and more agile planes. Biplanes were replaced by more streamlined monoplanes with a transition to all-metal frames made from high-strength lightweight aluminum alloys. One of the most striking planes to set an airspeed record was the Hughes H-1 racer built by Hughes Aircraft. The H-1 featured a glistening, completely smooth aluminum skin setting a world airspeed record on September 13, 1935 for a landplane of 354mph. Eighteen months later it set a new transcontinental speed record by flying non-stop from Los Angeles to New York City in a time of just under 7½ hours. Hughes's average speed over the duration of the flight was 322 mph. The H-1 was the last non-military plane to claim an air speed record.[9]

By the mid-1930s, inspired by planes such as the H-1 a new streamlined aircraft shape emerged, with tightly cowled multiple engines, retracting landing gear, variable-pitch fixed-wing propellers, and sleek aluminum alloy construction. Two of the most important examples of the new design, and making extensive use of aluminum, were the Douglas DC-2 introduced in 1934 and the DC-3 two years later in 1936.

The DC-2 could fly coast-to-coast faster than any passenger plane before. The larger DC-3 could seat up to 27 passengers in relative comfort. American Airlines was the first adopter of the DC-3 and they were followed by other major carriers of the time including United, TWA, Delta, and Eastern. By

1939, at least three-quarters of all air travelers were flying on DC-3 s. Statistics compiled by the U.S. Department of Commerce recorded that in 1932 there were 474,000 air passengers. That number increased to 1,102,000 in 1937 and to 1,176,858 one year later. Other statistics show that the number of passenger miles traveled in the United States increased 600% from 1936 to 1941. These numbers reflect the popularity of the Douglas DC-3 and the recognition of aluminum as the critical material for flight.[10]

During World War II, aluminum was an important material for many military applications with uses in battleships, radar, and even mess kits. In particular the metal was essential for the construction of the next generation of military aircraft. The Lockheed P-38 Lightning, introduced in July 1941, was the first American fighter to make extensive use of aluminum skin panels. It was also the first military airplane to fly faster than 400 mph on level flight.

Between July 1940 and August 1945, the United States produced a staggering 296,000 military aircraft. At the peak of production in 1944, American warplanes were rolling out of the manufacturing plants such as Boeing Field in Seattle at the rate of eleven every hour. More than half of these were made predominantly from high-strength low weight aluminum alloys.

Aluminum was a strategic material and domestic aluminum production capacity grew rapidly to keep up with the massive need for aircraft.

The demand for aluminum far exceeded the U.S. production capacity at the time. In 1939 annual production of primary aluminum (that obtained directly from the ore and not from recovery or recycling) was less than 164,000 tons. In 1942, New York City radio station WOR broadcast a show "Aluminum for Defense" to encourage Americans to donate their scrap metal for the war effort.[11] Recycling programs were encouraged, and "Tinfoil Drives" offered free movie tickets in exchange for balls of aluminum foil.[12]

By 1943 aluminum production had soared to over 900,000 tons. One of the consequences of the increased demand was a shift in the main center of aluminum production from the East Coast to my part of the country, the Pacific Northwest. Up until the outbreak of war not a pound of aluminum was produced in this region. By 1943, the Pacific Northwest accounted for over a quarter of all domestic production. The reason was simple. The region was one of the few locations in the country where surplus electric power was available, an outcome of our extensive hydroelectric generation capacity.

After World War II, the benefits of aluminum for airplane manufacture were obvious: it offered an excellent combination of low weight and high strength. While the composition of aluminum alloys continued to evolve with new formulations for even higher strength or better corrosion resistance or improved stability at high temperatures, the advantages of aluminum allowed postwar designers to build planes that were light, could carry heavy loads, use the least amount of fuel, and be impervious to corrosion ensuring the safety of the aircraft and its passengers.

In 1958 two de Havilland DH 106 Comets flying for the British Overseas Airways Corporation (BOAC) completed the first trans-Atlantic jet passenger service between the United States and Europe. One of the Comets departed from London, the other from New York. Flying at speeds up to 500 mph with a maximum cruising altitude of 40,000 feet the westbound flight completed its journey in 10 h and 22 min (including a refueling stop in Newfoundland). The eastbound flight with the advantage of some friendly tailwinds made the trip to London in what was a record-breaking time of 6 h and 11 min. Most airlines have this leg booked for 7 h! Propeller planes were taking over 14 h to cover the same routes.

The Comet with its mirrored aluminum fuselage, swept back wings, and four jet engines looked futuristic. Alistair Hodgson, curator of the de Havilland Aircraft Museum located in London Colney, in the UK sums it up nicely: "It [the Comet] was the Concorde of its day—it flew higher, faster, smoother than any other airline of that time and made everything else obsolete." Although the Comet ended up losing out to the Boeing 707 in the battle for airline adoption, aluminum featured prominently in both planes.

The Boeing 737, the best-selling jet commercial airliner, is 80% aluminum. The 737 has made short-haul and medium-range air travel for the masses a reality and has been manufactured continuously by Boeing since 1967. Close to 15,200 aircraft have been delivered with orders for over 4,700 yet to be fulfilled.[13] On average there are over 1,000 Boeing 737 s airborne at any given time, with two either departing and/or landing somewhere in the world every five seconds.

Aluminum alloys have revolutionized not only air travel but other forms of transportation, from trains and cars to high-speed superbikes. Aluminum is increasingly the material of choice for automakers. From mass-market vehicles like the Ford F-150 to luxury cars like Audi, Mercedes Benz, and Land Rover the use of aluminum is accelerating. On average, a car built in 2015 contained 397 lb of aluminum. Five years later this number is well on its way to exceeding 460 lb. At the same time the decrease in overall vehicle weight is boosting fuel economy, reducing emissions, increasing performance while maintaining or even improving safety.

Aluminum has played an important role in our exploration of outer space, where low weight coupled with high strength is even more essential than it is within the Earth's atmosphere.

On October 4, 1957, the Soviet Union launched the first satellite, the Sputnik 1, which was made from an aluminum alloy. The aluminum sphere was covered with a 1-mm thick aluminum-magnesium-titanium heat shield. All modern spacecraft are comprised of as much as 90% aluminum alloys. Aluminum alloys were used extensively on the Apollo 11 command and service modules. The command module housed the crew and the equipment necessary for re-entry into the Earth's atmosphere. Made of an aluminum honeycomb sandwiched between sheet aluminum alloy the command module was a conical pressure vessel with a maximum diameter of 3.9 m at its base and a height of 3.65 m. The service module carried most of the consumables (oxygen, water, helium, fuel cells, fuel) and the main propulsion system. The cylindrical service module had an outer skin of 2.5 cm thick aluminum honeycomb panels was attached to the back of the command module. Aluminum beams were used to divide the interior.

Anodized aluminum is used for coating handrails on the International Space Station (ISS). The handrails are used as anchors and lifelines keeping the astronauts from floating off into outer space when they venture outside for a spacewalk. On the ISS the handrails are anodized in a bright gold finish to stand out clearly against the surrounding structures.

Aluminum has a critical role in enabling the next phase of human space exploration—beyond the Moon. The Orion spacecraft—currently under development—is intended to allow travel to the asteroids and to Mars.[14] The manufacturer, Lockheed Martin, has chosen an aluminum–lithium alloy for Orion's main structural components. Lithium is the lightest of all the metallic elements, with only three protons in its nucleus. Only the gases hydrogen and helium have fewer. When aluminum is alloyed with lithium, a lithium atom displaces one aluminum atom from the crystal lattice while it still maintains the close-packed structure.

Commercial aluminum–lithium alloys contain up to 2.5% by weight of lithium. This might not seem like much, but every 1% of lithium added to replace aluminum reduces the density of the resulting alloy by 3% and increases the stiffness by 5%. The introduction of lithium creates a local distortion of the lattice in a similar way to the addition of copper to form duralumin. Except in the case of lithium, the introduced atom is smaller than the host. It takes extra energy to move a dislocation through the distorted region of the crystal making the resulting material stronger and more difficult to deform. It is also possible to heat treat aluminum–lithium alloys, as is done with aluminum-copper alloys. Under the appropriate conditions, tiny precipitates form. These further strengthen the metal by locally distorting the lattice impeding dislocation motion.

Producing aluminum is a highly energy intensive technology. Although not one of the top four industrial CO_2 emitters—those are cement, steel, ethylene, and ammonia—approximately 4% of all the electricity generated in the United States goes directly into aluminum production from alumina. Over 14,500 kWh of electricity is required to produce one ton of aluminum. This is considerably more electricity than the average annual consumption for a U.S. residential utility customer.[15]

In many countries, aluminum plants are located near hydroelectric power stations because of the requirement for massive amounts of low-cost electricity. Tomago Aluminium is one of Australasia's largest aluminum smelters and produces over 580,000 tons of aluminum every year; one quarter of Australia's primary aluminum. The Tomago smelter is located in New South Wales, which is one of the major regions in Australia generating hydroelectricity. Around 10% of the state's power supply is used to produce aluminum.

Nuclear and coal-fired power stations are also used in areas that do not have significant hydroelectric developments. Russian manufacturer Rusal, the world's second-largest aluminum producer currently uses a mix of hydroelectricity, nuclear power, and natural gas. By contrast, China's aluminum smelters, which produce half the world's output of aluminum, get 90% of their energy from carbon emission-heavy coal-fired power plants.

Since the Wright Brothers first took aluminum into flight the metal has become indispensable for not only flying us around the world, but also taking us into space. Even Boeing's innovative and disruptive 787 "Dreamliner", the first major commercial airplane to have a carbon fiber reinforced composite fuselage, composite wings, and to use composites for most of the other airframe components, still uses a significant amount of aluminum (about 20% of its total weight.)

Aluminum has gone from a sparingly used exotic precious metal to an indispensable commodity metal. Today the United States Geological Survey, part of the Department of the Interior, lists aluminum as a critical mineral because of the loss of domestic smelting capacity and the heavy reliance on bauxite imports, the mineral source of alumina.[16]

Notes

1. U.S. National Advisory Committee for Aeronautics, Annual Report, 6th (Washington, D.C., 1920), 52–53. Reported in Schatzberg, Eric (2003). Symbolic culture and technological change: The cultural history of aluminum as an industrial material *Enterprise & Society 4*, 226–271.

2. Byrne, John (2015) Outlook for Aluminum in the Commercial Airplane Market, https://www.aluminum.org/sites/default/files/10-21%20Gen%20Ses%20John%20Byrne.pdf Accessed 18 February 2019. The prediction is that by 2034 the global fleet will be 43,560 aircraft (double the number in 2014). Aluminum will continue to be the largest content in commercial airplane manufacturing, but it will increasingly need to compete with composites as their cost comes down and integrate with composites in future aerospace applications.

3. Pullman, Washington is named after George Pullman even though he never lived in Pullman nor were the famed railcars ever made here. Bolin Farr, a local homesteader, was a friend of George Pullman and named the city in his honor. https://wsu.edu/life/pullman/history/ Accessed 18 February 2019.

4. Pullman Progress 1859 Wood 1907 Steel 1933 Aluminum, The Pullman Company, Chicago, IL. Undated promotional brochure from The Pullman Company. Available online through the University of Chicago Library. https://www.lib.uchicago.edu/ead/pdf/century0062.pdf

5. Copper, silver, and gold are all fcc forming part of a block of eight metals with the same structure.

6. In a cube of aluminum one centimeter on each side there are about a billion dislocations that if placed end-to-end would encircle our planet 250 times.

7. Alfred Wilm gave the Dürener Metallwerke AG sole rights to his patents. The name Duralumin comes from a contraction of the company name and the parent metal. A history of duralumin is given by Olivier Hardouin Duparc (2005) Alfred Wilm and the beginnings of Duralumin Z. Metallkd. 96: 398–404. Another factor that affects the ease of dislocation motion is the metal bond strength. The bond strength in aluminum is 133 kJ/mol. For copper the value is 176 kJ/mol. The difference is reflected in the much higher melting point of copper than aluminum.

8. The weight of a typical engine of the time was 15 lb/hp. The Wright brothers developed an engine of 7 lb/hp using the lighter aluminum.

9. The SR-71A "Blackbird" developed by Lockheed holds the current manned air speed record of 2,193.167 mph set on 28 July 1976. https://www.museumofaviation.org/portfolio/sr-71a-blackbird/ Accessed 18 February 2019.

10. The United States Department of Transportation, Bureau of Transportation Statistics reported that US airlines carried an all-time high number of passengers during 2017—849.3 million systemwide. https://www.bts.dot.gov/newsroom/2017-annual-and-december-us-airline-traffic-data Accessed 18 February 2019.

11. An excerpt from the WOR program "Aluminum for Defense" can be heard at http://historymatters.gmu.edu/d/5158&title=%22Aluminum+for+Defense%22%3A+Rationing+at+Home+during+World+War+II Accessed 18 February 2019.

12. The Aluminum Association https://www.aluminum.org/aluminum-advantage/history-aluminum Accessed 18 February 2019. Does not include any effects of COVID-19.

13. These are data from Boeing up to January 2019. http://www.b737.org.uk/sales. htm Accessed 18 February 2019.
14. The first crewed flight is sceduled for 2024. https://www.lockheedmartin.com/ en-us/products/orion.html Accessed 18 February 2019.
15. In 2000 and 2001 Kaiser Aluminum Corporation in Spokane sold electricity it bought from the Bonneville Power Administration for $22 per megawatt hour back to the company for $500 per megawatt hour. It made more money reselling electricity than it did manufacturing aluminum. http://djcoregon.com/news/2000/12/22/kaiser-aluminum-corp-clo ses-spokane-plant-to-sell-electricity/ The gains for the company were, however, short term. Accessed 18 February 2019.
16. Fortier, Steen M, Nassar, Nedal T, Lederer, Graham W, Brainard, Jamie, Gambogi, Joseph, & McCullough, Erin A. (2018). Draft critical mineral list-Summary of methodology and background information-U.S. Geological Survey technical input document in response to Secretarial Order No. 3359: U.S. Geological Survey, Open-File Report 2018–1021, 15p.

References

1. Skillingberg, Michael & Green, John (2007). Aluminum applications in the rail industry. *Light Metal Age*. October: 1–5.
2. Hales, Thomas and others (2017). A formal proof of the Kepler Conjecture. *Forum of Mathematics Pi, 5,* e2.
3. Deville, H. (1859). *de l'Aluminum* Paris. The English translation is Deville H. (1933) *Aluminum,* R.A. Anderson trans. Sherwood.

11

Silicon—*The Material of Information*

Without silicon, the heart of the ubiquitous "silicon chip", the modern information age based on the computer, iPhones, wireless networks, and high-speed digital data transfer would not be possible. Imagine still having to consult a map to find directions rather than using satellite navigation or thumbing through a phonebook to find a phone number rather than having Siri do it for you. Or imagine that the only way to find information was going to the local reference library for a book rather than accessing the same information, and more, on-line in a fraction of a second. This transformation in how we access information and how we communicate with each other using social media has happened over a very short period of time (only a decade or two for some of these technologies; the first generation of iPhone was announced in January 2007, the Chinese social media site WeChat was released in 2011).

The history of silicon, as a material that has enabled all these technologies and many others goes back a little further than the iPhone, but it certainly does not have the long and rich history that characterizes some of the other materials in this book. However, the extent to which silicon has shaped our world cannot be overstated. Silicon is found in every single one of our electronic devices—washing machines, televisions, refrigerators, computers, toasters, Bluetooth speakers, and yes in smartphones.

To understand why silicon is so important and possibly irreplaceable as an active component in our rapidly expanding array of electronic devices it is necessary to look at its crystal structure, how it is purified and transformed

© The Author(s), under exclusive license to Springer Nature Switzerland AG 2021
M. G. Norton, *Ten Materials That Shaped Our World*,
https://doi.org/10.1007/978-3-030-75213-2_11

Fig. 11.1 A high-resolution electron microscope image of silicon. The bright circles are columns of silicon atoms with a spacing of only 0.136 nm (0.136 billionths of a meter). This is an excellent example of the resolution possible with state-of-the-art electron microscopes. (Reproduced with permission from SPIE and Professor Ki-Bum Kim. Source: Kim, Ki-Bum (2008) A novel technique for projection-type electron-beam lithography," *SPIE Newsroom* December 30)

into large single crystals, and how these single crystals are processed into transistors, which become embedded into integrated circuits. To give a perspective on just how enormous the scale of transistor manufacturing is there are over 56 billion transistors produced *per person per year*. Each of these transistors is cheaper than a single grain of rice [1].

So first let's consider how the atoms are arranged in a crystal of silicon. Silicon has the same crystal structure as diamond, which might not be surprising as silicon is just below carbon in group four of the Periodic Table of Elements. Each atom has four neighboring atoms forming a tiny tetrahedron (a triangular-based pyramid). This arrangement was determined in 1913 by father and son team of Cambridge physicists Sir William Henry Bragg and Sir William Laurence Bragg [2]. In a crystal of silicon the distance between adjacent atoms in each tetrahedron is only 0.2 nm, or 0.2 billionths of a meter. Each tetrahedron shares corners with four other tetrahedra, and so on, until we have an (almost) infinite array of corner-sharing tetrahedra. The regularity of this arrangement can be appreciated from the high-resolution transmission electron microscope image shown in Fig. 11.1. Each of the bright dots in the image represents a column consisting of just a handful of silicon atoms.

The silicon tetrahedra are held together by covalent bonds, which are shared electrons between neighboring atoms. All the electrons in a silicon atom are either tied up in these covalent bonds or held tightly by the silicon nucleus. The absence of unbound electrons is why silicon is not a good electrical conductor. There are no free electrons (also called conduction electrons) to move when an electric field is applied. Instead, silicon is a semiconductor.

Semiconductors are defined as materials that have an electrical conductivity at room temperature in between that of a metal (for example, aluminum which is how electricity crisscrosses the country through the power grid) and an insulator (for example, polyethylene, which was vital for insulating radar during World War II.) The reason that silicon is a semiconductor and not an insulator, like diamond, which has the same crystal structure, is that although the covalent bonds are strong, they are not *that* strong. (Semiconductors are the baby bear of materials: the bonding is not too strong, it is not too weak, it is just right.) It is possible to break electrons free from the covalent bonds— for example, by heating silicon. The hotter silicon gets the better electrical conductor it becomes.[1]

More importantly, the electrical conductivity of a semiconductor can be dramatically changed by the addition of miniscule amounts of deliberately added impurities called *dopants*. Although heat is one way to produce conduction electrons it is not very useful in terms of making tiny compact devices such as transistors. A better way is to add small amounts of dopant atoms to the silicon, which replace about one in every million silicon atoms and by doing so generate the precise number of mobile electrons that we need.

As an element in group four of the Periodic Table, each silicon atom has four outer electrons, which it shares with four other silicon atoms when it forms covalently bonded silicon tetrahedra. If a silicon atom is replaced with an atom from group five, for instance a phosphorus atom, it creates one conduction electron. The reason is simple. Phosphorus has five outer electrons. It uses four of these to form covalent bonds with four silicon atoms, leaving one extra electron that is not used for bonding. This electron is only very weakly tied to the phosphorus atom and can easily be forced to move when an electric field is applied.

On the other hand, if we replace one of the silicon atoms in a crystal with an element from group three, such as boron, we are inserting an atom that has three outer electrons. It uses these three to bond with three silicon atoms, but it can't complete the bond with the fourth silicon atom. The silicon has provided its electron, but the boron has no more to share. We say that the addition of the boron atom has created an "electron hole," or simply a "hole" in the covalent bond. Holes can move under the influence of an applied electric field, but they move in the opposite direction to electrons. So when a crystal of silicon is doped we either create an excess of conduction electrons or an excess of holes, depending on whether the dopant is an element from group three or group five. If there is an excess of conduction electrons the silicon is referred to as *n*-type (*n* for negative). On the other hand, an excess of holes results in a *p*-type (*p* for positive) semiconductor. Even though there is a miniscule amount of dopant atoms they have a disproportionate influence on the conductivity of the material.

Every semiconductor device from the simple diodes used during World War II to the most complex of today's computer chips uses *p*-type and *n*-type silicon. The trick is how these *p*-type and *n*-type regions are put together and the distance between them. The heart of any computer, smartphone, tablet, etc. is the field effect transistor or FET. The FET is conceptually quite simple. It is two *n*-type regions, the source and the drain, separated by a *p*-type region, the gate. The idea is that depending on the voltage applied to the gate we can either create a continuous electron path between the two *n*-type regions, which could be defined as a "1." Alternatively, the two *n*-type regions can be isolated from each other permitting no current to flow between them. This situation could be defined as a "0." By having 1's and 0's we have the elements of a binary system. And, what is really important, we can very quickly switch between the two states. By linking hundreds, thousands, millions, or billions, of FETs together we can perform an almost unlimited number of logical operations.

In addition to its almost ideal bond strength and its ability to be easily doped, silicon forms a perfectly uniform, tough, stable oxide layer on its surface when it is heated in air or steam. (This is very similar to the passivation of aluminum described in the previous chapter.) The oxide layer on silicon is composed of silica glass, which protects the underlying crystal surface. Although most of the solid elements react with oxygen, very few of their oxides adhere to the underlying surface and provide the level of protection that the silica layer does for silicon. For instance, iron oxide (rust!) readily forms on iron, but easily flakes off and doesn't protect the metal from continued rusting. It is the oxide layer on silicon that is used to isolate the three active regions of the transistor (the source, the gate, and the drain).

The process to turn a doped silicon crystal into an integrated circuit involves a great many steps, each of which requires a high degree of precision. But silicon is a material that was almost perfectly created by nature for manufacturing these devices. If we had the ability to design a new element that had every desirable feature necessary to mass produce inexpensive computer chips, it probably wouldn't be possible to improve on silicon. Enabled by silicon we have built a powerful, global, hyperconnected society.

The discovery of silicon is credited to the very important and influential Swedish scientist Jöns Jacob Berzelius, who also discovered and isolated the rare earth element cerium and the radioactive element thorium. Berzelius developed the now familiar and universally adopted chemical notation where elements are represented simply by one or two letters. For example, H is for hydrogen, O is for oxygen, Ce is for cerium, and Th is for thorium. In addition to these and many other pioneering contributions to chemistry, in 1824, Berzelius prepared a pure form of amorphous silicon (chemical symbol Si). The existence of the element had already been proposed by French nobleman and chemist Antoine Lavoisier in 1787 and given a name—"silicium"—by English chemist Sir Humphry Davy in 1808. The name silicon is derived from the Latin "silicis" meaning "flint," which as we saw in Chap. 2 is an abundant silicon-containing mineral. The ending "ium" was added because Davy thought that silicon would be a metal.[2] He turned out to be incorrect.[3] Subsequently, the name was changed by Scottish chemist and mineralogist Thomas Thomson, who correctly asserted that silicon was like boron and carbon: it is non metallic. In his four-volume chemistry text published in 1817 Thomson stated, referring to silicon: "There is not the smallest evidence for its metallic nature" [3].

The silicon prepared by Berzelius was amorphous, meaning that it was in the form of a glass, not the more usually encountered crystalline variety that is required to make silicon chips.

A crystalline form was eventually prepared in 1854 by French chemist Henri Étienne Sainte-Claire Deville. In the same year, Deville published a paper describing a method of producing aluminum that significantly lowered the cost of the metal.[4]

Silicon is plentiful in the Earth's crust, accounting for 27% by weight, which makes it the second most abundant terrestrial element after oxygen. The affinity of silicon for oxygen and the strength of the bond that forms between the two elements means that silicon is never found naturally in its elemental form.[5] It is always bound up in silicate minerals, such as quartz, and in aluminosilicates such as feldspar. Feldspars represent by far the most widespread mineral group in the Earth's crust, forming about 60% of all rocks. Feldspars are extensively used in the manufacture of glass. They are also found in many household products such as paint, plastic, and rubber.

The process to extract silicon from its mineral source requires multiple steps and, for two main reasons, is very energy intensive. Firstly, it is necessary to break the strong silicon-oxygen bond. And secondly, the electronics industry requires that the resulting product be extremely pure. "Electronic grade" silicon, the form that is used to make integrated circuits has impurity levels of less than one part per billion (ppb), making it among the purest materials ever made. One part per billion means that there is less than one impurity atom per billion silicon atoms. Almost no other synthetic material can claim this extraordinary level of purity.

The raw material for all silicon chips (we will use the term silicon chip rather than integrated circuit because integrated circuits can be based on non-silicon semiconductors) is sand. The sand found on most beaches generally has too many impurities to be useful, but relatively pure deposits are scattered around the world. In the United States—the world's largest producer of industrial sand—particularly high-quality deposits are mined from the St. Peter sandstone formation in Illinois and Missouri and from the Oriskany sandstone deposits in West Virginia and Pennsylvania. In these areas, the sand has a purity greater than 99% [4]. Sand is an important raw material not just for silicon but is a major constituent in the manufacture of glass and the formulation of concrete.

After it has been mined, the sand is shipped to a factory where it is placed in a furnace along with a source of carbon and heated to almost 1,500 °C. At this temperature, the sand reacts with the carbon to form carbon dioxide leaving a relatively impure form of silicon known in the industry as "metallurgical grade silicon," which has a purity of 98% and sells for about $2/kg.

Further purification requires converting the metallurgical-grade silicon into a highly flammable and very toxic gas called silane (SiH_4).[6] This conversion itself occurs over a number of steps but is very efficient (with almost 100% conversion) because at each step the by-products of the reaction are recycled. The high purity silane is then decomposed at 850 °C to produce electronic grade silicon, which is worth more than $100/kg. The decomposition reaction occurs on the surface of specially prepared single crystal rods of silicon. The electronic grade silicon grows out from the surface of the rod producing a feather-like structure consisting of a central shaft with a series of branches and sub-branches [5].

To create the temperatures high enough for the decomposition of silane an electric current is passed through the silicon rod, which is shaped like a large upside-down U. The resistance of the rod to the flowing current causes it to heat up, in much the same way as a tungsten filament is heated in an incandescent lamp. The need for large amounts of cheap electricity is one reason why many silicon manufacturers are located near hydroelectric plants such as Grand Coulee Dam in Washington State and the Liujiaxia Dam in China's

Gansu Province. This situation is very similar to the location of aluminum manufacturing plants that are also very energy demanding. Worldwide more than ten million kg (over 11 thousand tons) of electronic grade silicon is produced each year.

The next step in the process toward producing a silicon chip is to convert the electronic grade silicon, which has been ground into a powder, into one large single crystal weighing about 200 kg.

There are several ways that this transformation can be achieved but the almost universal method is to pull a crystal of silicon slowly from a melt using a method first developed by Jan Czochralski, a Polish chemist. The Czochralski method for growing crystals is yet another fine example of serendipity, a repeating theme in the history of materials. The story goes that in 1916 the then twenty-eight-year old Czochralski was working on the crystallization of metals, which happens when a liquid metal, such as iron, is poured into a mold to form cast iron. Czochralski was writing up his notes for the day while sitting next to a crucible containing molten tin (at a temperature of about 230 °C). Absentmindedly, instead of dipping his pen in the inkpot it went into the hot crucible containing the liquid tin. Czochralski quickly withdraw the pen from the molten metal whereupon he observed a thin thread of solidified metal hanging from the nib. His observation led to a new crystal growth technique that he published in 1917 and which bears his name, sometimes in industry called simply the "Cz" method.

The crystal growth process used by the electronics industry is a far cry from those early crude experiments performed by Czochralski. Silicon melts at over 1400 °C, much hotter than the melting temperature of tin, so it is a requirement to have a container that can hold this high temperature liquid without itself being dissolved. It is also necessary to shroud the silicon melt with an inert gas such as argon to prevent the silicon reacting with oxygen and reverting back into silica, which would basically take us back to where we started with sand, albeit in a much purer form. And finally, in order to ensure that we get as perfect a crystal as possible the pull rates—in other words, how fast Czochralski's pen was pulled from the melt—have to be very slow. Typically, the growth rates for silicon crystals are about 8 mm per day. The pulled crystal is called a *boule*, the French word for ball. Early crystals were shaped much more like balls than modern day crystals, which are pointy-tipped cylinders. When a silicon boule is grown using the Czochralski process it is usual that the boule will be doped to make it either *n*-type or *p*-type. This is achieved by adding the appropriate dopant in a carefully controlled amount directly to the silicon melt. For example, to make *p*-type boules high-purity powdered boron might be added. The boules are worth about $400/kg.

Two people who were particularly important in producing appropriately pure and defect-free silicon crystal boules for the production of silicon chips were Gordon Teal of Bell Telephone Laboratories and William Dash of General Electric Research Laboratories. In May 1952, Teal and his colleague Ernie Buehler reported at the annual meeting of the American Physical Society held in Washington, D.C., that they had successfully produced large single crystals of silicon using the Czochralski method [6]. A persistent problem with the silicon boules being grown at the time in Bell Laboratories was the presence of dislocations. The importance of dislocations in influencing the mechanical properties of metals was something that we noted with both wrought iron and aluminum; we generally want to reduce their mobility. In single crystal silicon we want to avoid dislocations all together because they adversely affect the electronic properties of the final integrated circuit. Dash, a crystal grower at General Electric, used a novel variation of the Czochralski process to produce silicon crystals completely free from dislocations [7]. Early on as the crystal was being withdrawn from the melt, Dash rapidly increased the pulling rate to form a very narrow "neck." Any dislocations that are present in the growing crystal glide to the surface of the neck and, essentially, disappear. Dislocations increase the internal energy of the crystal. Given the right encouragement, in this case temperature, they will annihilate either with each other or with the silicon surface. Once all the dislocations had been removed, Dash slowed the growth rate producing a dislocation-free silicon of the desired size.

As the industry became better at controlling the crystal growth process the diameter of the boules that could be grown increased. In the 1960s, at the very beginning of the modern electronics industry based on silicon, the boules were sliced to produce wafers that had a diameter of 2 in. Now twelve-inch diameter wafers are the industry standard and there is a push towards production of eighteen-inch wafers. The larger size boule lowers the cost per wafer. Cut and polished twelve-inch wafers sell for about $1,300/kg.

After each step of a very demanding process the value of the product has significantly increased.

What makes silicon chips so inexpensive is that they can be mass produced, which is why there is the drive to make increasingly larger boules. In 1960 output was less than one transistor per person per year. Today it is well over 50 billion per person. The industry is efficient because it employs economies of scale. Making one chip essentially costs as much, and takes the same time, as making several hundred. As an illustration of the efficiency of the process we compared the cost of a transistor to that of a single grain of rice. Another comparison is that the cost to print each letter in this book is less than the cost of each transistor that was used in the laptop that created each letter [8].

Before we look at how silicon became an indispensable part of our technology, let's briefly examine what the electronics industry was like before the advent of the silicon chip. From the early 1900s through to the 1960s the vacuum tube was the most important electronic device. From the outside a typical vacuum tube from the 1920s resembles an incandescent light bulb. It is simply a transparent glass bulb surrounding a vacuum. Within the bulb are two separated metal wires called electrodes. A light bulb contains just a single continuous metal filament that glows when it is heated.

What was first noted by Thomas Alva Edison—who else!—and subsequently named the Edison Effect, is that in a vacuum an electric current can jump from one metal electrode to another. The electrons can pass right through the vacuum. Edison made this observation in 1883, but it took over twenty years before anyone did anything useful with it [9]. Then, in 1904 British electrical engineer and physicist Sir John Ambrose Fleming made a vacuum tube device that operated like a valve. It allowed an electric current to flow in only one direction, but not in the other. These devices are known as diodes. An important application for Fleming's diodes was in radios, which required direct current (DC) for their operation, but the input supply—from the outlet—was alternating current (AC). By allowing current flow in one direction only, AC was converted to DC, which powered the radio. AC to DC conversion is called rectification: a diode is a rectifier.

In 1906 American engineer Lee De Forest modified Fleming's diode design by placing a metal grid, in addition to the two electrodes, inside the glass bulb. By varying the voltage that was applied to the metal grid De Forest could control the flow of electrical current through the tube. This device was a *tri*ode (originally called an "audion" by De Forest) because there were three electrodes (the original two of Fleming's diode and now the metal grid). The most important feature of the triode was that it could be used to take a weak electrical signal and amplify it into a larger one through control of the grid voltage. The engineers at Bell Telephone Laboratories quickly recognized the potential of the triode for telephony and purchased the patent rights from De Forest. In 1913, American Telephone and Telegraph Company (AT&T) installed audions to boost voice signals as they crossed the 3,400 miles of wires connecting one coast to the other. (Without amplification calls were limited to a distance of 800 miles.)

On January 25, 1915, Alexander Graham Bell, sitting in New York, spoke into his telephone the words he had first used on March 10, 1876: "Mr. Watson, come here. I want you." Thomas A. Watson, sitting this time in San

Francisco (rather than in the same Boston laboratory as Bell) replied, "It will take me five days to get there now!".[7]

The audion was pivotal in the development of long-distance telephone communication. It would also revolutionize the first mass communication device, the radio. Amplification provided the power to drive loudspeakers, making listening to the radio a family event.

Another important use of radios was for maritime navigation. Transmitting and receiving Morse code signals allowed ships to communicate over the vast expanses of ocean, which reduced their isolation and, as a result, saved thousands of lives. One example that highlighted the value of radio to ocean vessels is the sinking of the R.M.S. *Titanic*. After colliding with an iceberg that tore a 300-foot gash in its steel hull, as the ship sank towards its final resting place in the North Atlantic Ocean senior wireless operator John George "Jack" Phillips sent out the emergency distress signal. The R.M.S. *Carpathia* picked up the S.O.S call and was able to help over 700 of the stranded passengers. Phillips was born in Godalming, Surrey: my hometown. A park adjacent to the River Wey was created as a memorial and bears his name for "still sending out distress calls to the last."[8] Unfortunately, the Titanic was not equipped with the most up to date radio equipment that had recently become available and was, instead, using the older Marconi radio system. The *New York Times* reported on May 2: "Sixteen hundred lives were lost that might have been saved if wireless communication had been what it should have been."[9]

The radio enabled information to be shared with hundreds, thousands, and eventually millions of people across the country and around the world. By 1926, the Bell Laboratories tube shop at 395 Hudson Street in New York City was turning out vacuum tubes for televisions, radios, public address systems, and telephone repeaters for the transcontinental telephone system. By the end of 1930, vacuum tubes were demonstrating lifetimes up to 20,000 h (over 2 years) and Bell Systems circuits alone were using over a quarter-million tubes [10].

In 1943 construction of the Electronic Numerical Integrator and Computer (ENIAC) began, in response to the need of the U.S. Army for calculating complex wartime ballistics tables. Speedy calculations were needed to inform a soldier just what settings a particular piece of artillery needed under a specified set of conditions. Although slow and cumbersome by today's standards the ENIAC could perform numerical computations a thousand times quicker than the mechanical calculators that were used at the time. This early computer could sum 5,000 numbers in a single second. Unfortunately, the ENIAC was not ready until the war was over—it was completed in November 1945. But over the decade since the end of the war until the ENIAC was struck by lightning in 1955, the first all-electronic computer is estimated to have run more calculations than all humankind had done up to that point.

Although the full scale of the impact that computers would have on society surely could have only been imagined by science fiction writers, John Mauchly, one of ENIAC's inventors, foresaw one vision of the future: "better weather-predicting—months ahead—better airplanes, gas turbines, micro-wave radio tubes, television, prime movers, projectiles operating at supersonic speeds carrying cargoes in peace and even more and better accuracies in studying the movement of the planets." Mauchly's co-inventor, J. Presper Eckert, Jr., went on to say: "The old era is going, the new one of electronic speed is on the way" [11].

Despite the great technological accomplishments made using vacuum tubes, their limitations soon became all too apparent. Even though great efforts had been made to increase reliability and improve their performance in many applications, they were bulky. There was no way around it; vacuum tubes took up a lot of space. The ENIAC contained 17,468 vacuum tubes.

On average one vacuum tube failed every other day and had to be manu-ally replaced. Furthermore, the ENIAC was heavy weighing in at over 30 tons. The other problem that was impossible to avoid was that vacuum tubes needed to be hot to work: they had to be heated, which was inefficient. There was a need for a new, revolutionary, approach. So, while both Mauchly and Eckert were correct in thinking of a high-speed electronic future enabled with computer technology, that future was not with vacuum tubes. It was with semiconductors. Specifically, it was with silicon.

In 1946, just a year after the completion of ENIAC a team at Bell Labo-ratories under the direction of William Shockley began working on crystal diodes as possible replacements for the bulky and slow vacuum tubes. They decided that they would focus their research efforts on the simplest two semiconductor materials: silicon and germanium. Even though, in hindsight, Shockley's decision might seem to have been a no-brainer, at the time the most important semiconductor materials were selenium (another element discovered by Jöns Jacob Berzelius, in 1817) and the compound copper oxide. Both these materials were difficult to work with.

A challenge when working with selenium is that although it is an element it exists in several different crystal structures and it can change between these structures when it is heated or cooled. This phenomenon—known as allotropy—is not uncommon among the elements, but it made working with selenium complicated because it became necessary to control the range

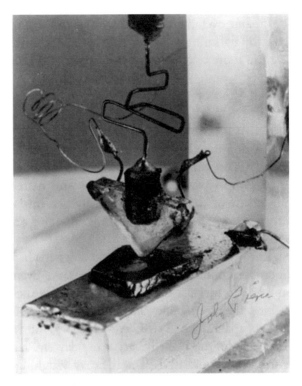

Fig. 11.2 Bell Laboratories' first transistor. This simple looking device shaped the modern world and eventually led to Facebook, Google, Artificial Intelligence, and the Internet of Things. (Source: Courtesy of the Computer History Museum.)

of temperatures over which the material was operated. On the other hand, the composition of copper oxide can change based on how much oxygen is present in the material. This phenomenon—known as non-stoichiometry—is not uncommon among oxides, but it means that the properties can be difficult to control and can vary from one sample to the next. Furthermore, the rectifiers that had been made using copper oxide were just not very efficient. In contrast to copper oxide, silicon and germanium are both elements and so are compositionally very simple: they consist of just one type of atom. And in contrast to selenium, they also have one of the simplest crystal structures.

Because of the direction taken by Shockley and his team, in December 1947 a scientific breakthrough occurred that was as big as anything that had ever happened before in the world of materials: Shockley, John Bardeen, and Walter Brattain constructed and tested the first transistor, which is shown in Fig. 11.2. It consists of a small crystal of germanium on a metal block. Pushing down on the germanium is a plastic triangle that has a thin strip of gold placed on each of the opposite sides. At the tip of the triangle these gold strips are separated by a distance of only 0.004 cm (about half the width of a human hair) and make electrical contact, a point contact, with the germanium. Brattain reported in his notebook that when an electric field was applied to the metal contacts they observed a small "power gain": the output power being greater than the input power. This phenomenon became known as the "transistor effect." The three researchers proudly demonstrated their invention to the management at the laboratory on December 23, 1947 and the following day, on Christmas Eve, they showed that the device would oscillate. Oscillation was the unambiguous proof of the existence of power gain. Walter Brattain's notebook stated: "By measurements at a fixed frequency it was determined that the power gain was the order of a factor of 18 or greater." This demonstration tipped the balance, everyone agreed that a very significant discovery had been made. The point contact transistor certainly looks primitive by today's standards, but this simple device created a revolution in the electronics industry that would change forever the way that we communicate, share information, and interact with each other.

The first public demonstration of the germanium transistor was on June 30, 1948, in the Bell Laboratories auditorium at their West Street location in New York City. The conference was well attended. Pictures of the event show a packed house, but the press showed little interest. The *New York Herald Tribune* article the following day read: "However, aside from the fact that a transistor radio works instantly without waiting to warm up [vacuum tubes had to be heated], company experts agreed that the spectacular aspects of the device are more technical than popular." That understatement matches that of William Orton, the president of Western Union, who in 1876 decided not to pay Alexander Graham Bell $100,000 ($2.2 million today) for the patent rights to the telephone, declaring the apparatus little more than a toy. Orton wrote to Bell: "After careful consideration, while it is a very interesting novelty, we have come to the conclusion that it has no commercial possibilities." Two years later Orton said he would consider the patent a bargain at $25 million (half a billion dollars today)!

The trade press was more favorable towards the newfangled germanium transistor. The September 1948 issue of *Electronics* magazine made the transistor its cover story. The accompanying article stated: "Because of its unique properties, the Transistor is destined to have far-reaching effects on the technology of electronics and will undoubtedly replace conventional electron [vacuum] tubes in a wide range of applications." In less than sixty years the transistor would become the dominant factor in electronics technology and integrated into every aspect of our daily lives.

As unfortunately so often occurs in the electronics industry—think Apple and Samsung today!—there was a dispute over who should be credited as the inventor of the transistor. Shockley believed it was his idea alone and that he should be the sole inventor named on the patent. He called Bardeen then Brattain into his office and told them each so. Bardeen apparently stormed out, while Brattain said 'Oh, hell, Shockley, there's enough glory in this for everyone!" [12] At the end of the day, the original point contact transistor made out of germanium was based on the work of Bardeen and Brattain and they were awarded U.S. Patent 2,524,035 in October 1950, while Shockley received U.S. Patent 2,569,347 for an improved transistor design called the junction transistor. Brattain proved to be correct and there was glory enough for all, with the three transistor pioneers sharing the 1956 Nobel Prize in Physics. (Bardeen got a second Nobel Prize in 1972 for his work on super-conductivity where he explained why superconductors superconduct—it is known as the Bardeen-Cooper-Schrieffer, or BCS, theory.)

In 1955, as transistor sales reached 4 million units, William Shockley left Bell Laboratories and started his own company, Shockley Semicon-ductor Laboratories in Santa Clara Valley, California. He recruited twelve outstanding young scientists as the nucleus of the new company. Unfortu-nately, Shockley had the ability to rub people the wrong way, as he had done in those early days with Bardeen and Brattain. Within a year eight of the initial twelve employees left Shockley Semiconductor Laboratories to start out on their own. These so-called "Traitorous Eight" included Gordon Moore, Jean Hoerni, and, probably most importantly, Robert Noyce (who went on to found Intel, whose chips are in over 80% of all computers shipped today). Although Shockley Semiconductor Laboratory was not successful their facility launched the high-tech magnet that became known as Silicon Valley, which became home to some of the biggest and most influential tech-nology companies in the world. The original Shockley building at 391 San Antonio Road, Mountain View became a produce market in 2006 and was eventually demolished.

Using $3,500 of their own money, this group of eight defectors from Shockley Semiconductor Laboratories developed a process to mass-produce transistors on a single wafer of silicon. This new process was an impor-tant technological breakthrough because up till that time transistors were manufactured individually, one at a time. Fairchild Camera and Instruments Corporation saw the potential of this new process and invested $1.5 million and on October 1, 1957, a new company Fairchild Semiconductor was born. Its first order was from the computer giant IBM for 100 transistors at $150 a piece. The wonderful part of the story is that the order was shipped in a Brillo

Pad carton picked up at a local supermarket by Jay Last (an MIT physicist and one of the "Eight"). NASA was also a customer of Fairchild Semiconductor and used their silicon transistors for the Apollo space program.

In 1947, when the team at Bell Laboratories were demonstrating the world's first transistor, U.S. production of germanium, the active material, was just a few pounds a year. By 1960, when integrated circuits were replacing individual transistors domestic germanium production exceeded 45 tons, all of it destined for the burgeoning electronics industry. Up until 1965 sales of germanium transistors outpaced those made of silicon. In that year 334 million germanium units were sold compared to 273 million made of silicon. The following year it switched. Silicon became the dominant material (481 million silicon transistors were sold exceeding the 369 germanium transistors) a position it has not relinquished and possibly never will.[10] So, in the space of two decades silicon won out as the semiconductor material of choice and became the material through which we get so much of our of information and misinformation.

There are a number of reasons why silicon eventually dominated germanium and why it is very difficult to see it being replaced by any other material in the near future. Firstly, silicon is more abundant than germanium. As noted earlier in this chapter, approximately 27% of the Earth's crust is made up of silicon. On the other hand, germanium accounts for only a miserly 0.00014%. This is less than rare-earth elements such as lanthanum, neodymium, and praseodymium and less even than the noble gas argon! Consequently, germanium is considerably more expensive than silicon. The only known significant source of germanium in the Western Hemisphere is in the tri-state area of Kansas, Missouri, and Oklahoma, where it is extracted from lead and zinc ores. But even with cost issues aside, silicon offers a big advantage when it comes to fabricating integrated circuits.

The invention of the integrated circuit, like the transistor itself, also has a disputed history. On February 6, 1959, Jack Kilby of Texas Instruments in Dallas filed a patent for an integrated circuit using germanium. This patent was eventually awarded on June 30, 1964, as U.S. Patent 3,138,743. On July 30, 1959, Fairchild filed a patent for a "monolithic" integrated circuit using silicon, which was based on Robert Noyce's work. This patent was awarded less than two years after filing on April 25, 1961, as U.S. Patent 2,981,877. An important part of both patents is the "use of a stable protective insulating layer that can be selectively etched allowing connections to be made to different parts of the circuit." By a series of masking and etching steps the pattern of the insulating layer can be varied according to the role of the chip. In silicon, this "stable protective insulating layer" is silica glass, which protects the underlying surface. No other semiconductor forms such a protective oxide. For instance, germanium oxide (GeO_2) is unstable making it unsuitable as an insulating layer. Without the stable oxide layer formed on silicon, integrated circuits, and thereby, computers would be much more expensive because they would be considerably more difficult and time consuming to fabricate.

The patent battle between Texas Instruments and Fairchild lasted for a decade. Finally, the courts came to a conclusion. Fairchild had the rights to an interconnection technique putting all of the components on a chip of silicon and connecting them with metal lines printed on an oxide layer, but Kilby had invented the integrated circuit. What is undisputed is that the first manufacturer of silicon chips was Fairchild Semiconductor and in 1962 they opened a facility in South Portland, Maine for the manufacture, test, and assembly of transistors for among other things, radios.

Jean Hoerni, a colleague of Noyce's, improved on the original Fairchild design and placed the three active components of the transistor—the emitter, the base, and the collector—onto a single flat plane creating a planar integrated circuit, which contained many transistors together with other circuit components all on a single wafer of silicon. This invention was critical because it not only improved the reliability of silicon transistors but also provided advantages in ease of processing and lower cost. The planer integrated circuit became the modern integrated circuit or silicon chip. The processes proposed by Hoerni over fifty years ago are essentially what we are using today to make the devices that go into every smartphone, laptop computer, and tablet.

The level of integration—the number of components that can be placed on an individual chip—has followed an observation made by Gordon Moore in 1965. Moore's Law, as it has become known, predicts that the number of components such as transistors that can fit on a silicon chip will double every year. In 1965 that number was 60. A decade later the number of integrated components had shot up to 60,000 validating Moore's law and establishing it as a guiding principle for semiconductor manufacturers. Since the 1970s the trend has deviated slightly from Moore's original predictions. Over the period 1971 to 2016 transistor counts have doubled every two years, rather than every year. The Intel 486 chip broke the 1 million-transistor mark in 1989. The computer chips used for gaming such as those in the Xbox One contain over 5 billion transistors, and the most advanced chips such as the SPARC M7 processor that is used in cloud computing have reached the staggering 10 billion mark.

Each day billions of people connect to the Internet through an estimated 30 billion devices. While the personal computer remains at the center of this evolving Internet of Things, Internet connectivity is now an integral part of cars, it is embedded in fitness equipment, used in factory robots and vending machines. Almost any device that has an on and off switch can be connected to the Internet. Every one of these devices relies on silicon and operates by the movement of electrons between tiny regions on a transistor that are much smaller than a virus. Our smarter, interconnected world has changed how we live, work, and communicate.

As we think to the future, will the history of silicon go the same way as that of flint: a once transformative material that lives on only in museums. For the moment silicon is the most important material for transistors, but transistors have been made from materials others than silicon. In 1998, Cees Dekker and colleagues at Delft University of Technology in the Netherlands made the first carbon nanotube transistor, which formed the basic building block of the carbon nanotube computer made in 2013 by Subhasish Mitra and Philip Wong of Stanford University in California [13, 14]. Although the carbon nanotube computer is only comprised of a paltry 178 transistors compared to the billions of transistors in a state-of-the-art silicon chip, it could represent a revolution in computing similar to the switch from vacuum tubes to silicon fifty years ago. In principle, carbon nanotubes allow the possibility of faster and more efficient transistors, but overall the carbon nanotube computer is much slower than current silicon computers because it has not been optimized. A major challenge with carbon nanotubes is the expense of scaling up production of a carbon nanotube computer as it is not clear how the benefits of integration that allowed silicon to triumph over germanium can be applied to assembling the individual nanotubes.

Before we write silicon off as the enabler of our future in information storage and communication, it is the element at the heart of one variety of quantum computer. Keiji Ono of the Advanced Device Laboratory of RIKEN in Japan explains: "Companies like IBM and Google are developing quantum computers that use superconductors. In contrast, we are attempting to develop a quantum computer based on the silicon manufacturing techniques currently used to make computers and smart phones. The advantage of this approach is that it can leverage existing industrial knowledge and technology" [15] .

Quantum computers although still in their infancy can solve problems much faster than conventional computers and tackle problems that are too complex for our existing technology. Yet there may be downsides to this ultrafast next generation of computers. The United States National Institute of Standards and Technology has predicted that within fifteen years the first quantum computer will emerge to defeat the most prevalent forms of encryption, exposing our most sensitive data and confidential information. On the other hand, quantum computers may create unbreakable protection for communication networks in the future. In either case advanced materials will be the key to enabling advanced information technology.

Notes

1. The increase in electrical conductivity of silicon when it gets hot is one of the reasons that most electronic devices are designed to provide cooling either by using a fan or controlling heat flow through, for example, the casing. In laptops such as the MacBook, excess heat is conducted through the aluminum casing.
2. Most metals named after 1800 follow the general convention that their names end with "-ium.".
3. Metals are defined by Merriam-Webster in terms of both their typical properties and their reactivity: any of various opaque, fusible, ductile, and typically lustrous substances that are good conductors of electricity and heat, form cations by loss of electrons, and yield basic oxides and hydroxides. https://www.merriam-webster.com/dictionary/metal
4. Three years later Deville discovered the compound silicon nitride; a synthetic ceramic that was of great interest in the 1970s for a lightweight high temperature gas turbine engine. Despite significant public and private financial investments the technology never materialized.
5. Bond Dissociation Energies have been compiled by Yu-Rao Luo and are available at http://staff.ustc.edu.cn/~luo971/2010-91-CRC-BDEs-Tables.pdf Accessed 14 February 2019. The Si–O bond strength is 800 kJ/mol, this can be compared with the bond strengths of other abundant elements to oxygen, e.g., O–Al = 502 kJ/mol, O-Fe = 407 kJ/mol, O-Ca = 383 kJ/mol.
6. Although major incidents involving silane are rare because of the safety procedures in place there are occasional leaks that lead to explosions. The most recent seems to be at the SunEdison plant in Pasadena, California in 2012. Four people were injured, but none seriously.
7. Alexander Graham Bell's notebook in which he uttered this famous directive to Thomas Watson is housed in the Library of Congress and available at https://www.loc.gov/item/today-in-history/march-10/
8. *The New York Times*, October 18, 1927.
9. New wireless rules America's demand *The New York Times*, May 2, 1912.
10. *Electronic Industries Association Yearbook*, 1967.

References

1. Qualman Darrin (2017). https://www.darrinqualman.com/global-production-transistors/ Data is from 2015 and shows the enormous increase in transistor production per capita since 1955.
2. Bragg, W. H. & Bragg, W. L. (1913), The structure of the diamond. *Proceedings of the Royal Society A: Mathematical, Physical and Engineering Sciences* 89: 277. W.H. Bragg was Cavendish Professor of Physics at the University of Leeds, W.L. Bragg had finished his BA and was at Trinity College, Cambridge. The diamonds used in the study were lent to the Braggs by the Mineralogical Laboratory at Cambridge. They used the recently developed technique of X-ray diffraction to solve the crystal structure.
3. Thomson, Thomas (1817). *A System of Chemistry in Four Volumes*, (5th edition. *1*, 252). London: Baldwin, Cradock, and Joy.
4. Davis, J. G. (2014). St. Peter sandstone mineral resource evaluation, Missouri, USA, in: *Proceedings of the 48th Annual Forum on the Geology of Industrial Minerals*, Phoenix, Arizona, April 30—May 4, 2012. Arizona Geological Survey Special Paper #9, Chapter 6, p. 1–7. The proceedings editor was F.M. Conway.
5. Fancher, R. W., Watkins, C. M., Norton, M. G., Bahr, D. F., & Osborne, E. W. (2001). Grain growth and mechanical properties in bulk polycrystalline silicon. *Journal of Materials Science, 36*, 5441–5446.
6. Teal G. K. & Buehler, E. (1952). Growth of silicon single crystals and of single crystal silicon p-n junctions. *Physical Review, 87*, 190.
7. Dash, W. C. (1959). Growth of silicon crystals free from dislocations. *Journal of Applied Physics, 30*, 459–474.
8. Browne, John (2013). *Seven Elements that have Changed the World* (p. 211). London: Phoenix, London. Uses the same analogy to illustrate the cost of transistors.
9. Edison, Thomas A. *US 307031.* Patent issued October 21, 1884.
10. Smits, F. M. (1985). *A History of Engineering and Science in the Bell System: Electronics Technology (1925–1975)*. Indianapolis: AT&T Bell Laboratories. Volume edited by F.M. Smits.
11. Electronic computer flashes answers, may speed engineering. *The New York Times*, February 15, 1946.
12. van Dulken, Stephen (2007). *Inventing the 20th Century: 100 Inventions that Shaped the World* (p. 122). New York: Barnes & Noble.
13. Trans, S. J., Verschueren, A. R. M., & Dekker, C. (1998). Room-temperature transistor based on a single carbon nanotube. *Nature, 393*, 49–52.
14. Shulaker, M. M., Hills, G., Patil, N., Wei, H., Chen, H-Y., Wong, H. S. P., & Mitra, S. (2013). Carbon nanotube computer. *Nature, 501*, 526–530.
15. Ono, K., Giavaras, G., Tanamoto, T., Ohguro, T., Hu, X., & Nori, F. (2017). Hole spin resonance and spin-orbit coupling in a silicon metal-oxide-semiconductor field-effect transistor. *Physical Review Letters, 119*, 156802.

12

Conclusion

My first computer, an Apple Macintosh Classic, contained about 1 million transistors, each made out of silicon. Ten years later the transistor count on a silicon chip had passed 100 million. It was over 10 billion by the end of the next decade. The exponential increase in the number of transistors in a single circuit known as Moore's Law, represents amazing developments in materials and how those materials are processed.[1] The rapid growth we have witnessed in many areas of technology has not been limited to electronics. Another example is in flight. At the dawn of the twentieth century humans had not developed the technology necessary to enable powered flight. Midway through the century the maximum distance flown by a non-commercial aircraft without refueling had reached almost 20,000 km (12,500 miles). At the end of the century that number was just under 41,000 km (25,500 miles). Both the rapid progress in electronics and in flight have been enabled by innovations in materials and design. The exponential growth rates observed over the last few decades provide a promise of more exciting technological advances in the future.

The evolution of silicon chips allowing the continuation of Moore's law over the past fifty-five years has been due to advances in shrinking the size of individual transistors. The smallest transistors in production are now only 5 nm—about the distance between 20 silicon atoms. For comparison, a typical virus is 100 nm. Even many proteins are larger than the features we can create on a silicon chip. A question is: how small is it possible to get with silicon? Intel is already planning features down to 1.4 nm, equivalent to only

M. G. Norton, *Ten Materials That Shaped Our World*, https://doi.org/10.1007/978-3-030-75213-2_12

12 silicon atoms across, by 2029. As some researchers are looking at ways to make smaller and smaller transistors to keep pace with the projections of Moore's law others are looking at materials to replace silicon.

Compound semiconductors such as indium gallium arsenide (InGaAs, pronounced "in gas") show promising performance for future high-speed applications and several universities and companies have already made InGaAs transistors.

A challenge with many of the compound semiconductors, including InGaAs, is that they involve some of the heavy and rarest elements. These are the elements that were formed after the Big Bang. Indium formed as neutron stars merged. Gallium and arsenic were created in powerful stellar explosions. These rare cosmic events account for the small concentrations of heavier elements in the Earth's crust. For instance, indium makes up only 0.21 part per million of the crust (less than many of the so-called rare earth elements).

A feature of many of the products of our modern world is our ability to integrate multiple materials into a single device. An example is the iPhone. The screen contains more than a dozen elements. The main active component is a thin film of indium tin oxide (ITO), which is an unusual material in that it is both transparent and a good electrical conductor. Those two properties rarely overlap in one material. The battery is made up of more than half a dozen elements among those are lithium and cobalt. There are over two dozen elements that comprise the electronic circuitry and components inside the smartphone. These include the more abundant metals such as copper as well as rare earth elements including gadolinium and dysprosium. Gadolinium is a component in the magnets inside the speaker and microphone. Dysprosium is part of the vibration unit that warns us that a text message or e-mail has just arrived in our inbox. The casing is made of lightweight magnesium alloys that are even lighter than those used in airplanes and space craft. And the plastic coating on the outside of the casing includes bromine-containing fire retardants. All of these materials combined weigh less than 200 g.

Unlike our earliest ancestors fashioning tools from flint and figurines from clay, materials that were abundant and widespread we have become increasingly concerned about where our materials come from, the environmental and social impacts related to their extraction, and what happens at the end of their useful life. One example that illustrates these concerns is tantalum a metal that is ubiquitous in microcapacitors that are used to reliably store large amounts of charge in a very small volume. These devices are critical to the operation of all mobile electronics including smartphones, tablets, and laptops where we want a combination of lightness with a sleek profile. Tantalum oxide (Ta_2O_5) is also used in glass lenses to make lighter camera lenses that produce a very bright image. Another requirement for smartphones.

The source of tantalum is the mineral tantalite also known by its nickname, coltan. About two-thirds of coltan is used in the fabrication of capacitors. Tantalum is one of the 3TG minerals—tungsten, tantalum, tin, and gold. These are the so-called conflict minerals. Tantalum is unique among this group as it is the only conflict mineral whose primary production is in the Great Lakes region of Africa, namely Rwanda, the Democratic Republic of Congo (DRC, Kinshasa), and to a lesser extent Burundi. Last year over 1,000 tons of tantalum came from this region with 740 tons coming from the DRC alone.[2] Profits from coltan mining and the other minerals that can be processed into one of the 3TG metals have been used as a source of funding for armed militias associated with vicious civil conflicts in the region. Consumer electronics, particularly the smartphone, have been highlighted as connected to the demand for these minerals from within conflict areas of the DRC [1].

Conflict over natural resources, including minerals, is not new. As we saw in Chap. 5, the history of gold is just one example of a material who acquisition has fueled brutal conquests and enslaved indigenous peoples. Diamond and silver would be others. Over the past several decades increasing scrutiny has been paid to revenues from natural resources that have fueled conflicts in developing countries. For instance, in 2010, the Dodd-Frank Act was adopted requiring companies that report to the United States Securities and Exchange Commission (SEC) to disclose the use of conflict minerals in their products.[3] In 2019, Apple removed eighteen smelters and refiners for flouting the conflict mineral code of conduct.[4]

The DRC is also the dominant miner of cobalt accounting for over 70% of global annual production. Why is cobalt important? Every lithium-ion battery in our smartphones, tablets, laptops, and electric vehicles requires cobalt to recharge. Cobalt combined with lithium and oxygen as the compound with formula $LiCoO_2$ forms one half of every lithium-ion rechargeable battery. The conditions for cobalt mining in the DRC are horrific. Children, as young as twelve, are forced to work as diggers in the mines where injury and even death are frequent occurrences [2].

Worldwide annual production of cobalt is currently around 140,000 tons. An analysis by Elsa Olivetti at the Massachusetts Institute of Technology and her colleagues estimate that by 2030 global demand for cobalt could reach between 235,000 and 430,000 tons. Although cobalt reserves could support this growth, current refining capacity is not sufficient [3].

The European Union and the United States have identified both cobalt and tantalum as critical minerals. Critical materials are those that are important to industry and defense, enable modern technology, such as our smartphone example, and are closely linked to clean energy technology.

The idea of critical materials is not new. As we noted in Chap. 8 United States President Franklin Roosevelt called rubber a "strategic and critical material" and created the Rubber Reserve Company in 1940 to stockpile reserves of natural rubber and regulate the production of synthetic rubber. Rubber was essential for the war effort and the Allies were concerned about losing access to natural sources of rubber and at the time not having the ability to produce sufficient quantities of synthetic rubber.

A recently issued Executive Order on a Federal Strategy to Ensure Secure and Reliable Supplies of Critical Minerals prioritizes developing recycling and reprocessing technologies, identifying substitute materials, and developing new and improved processes for critical mineral extraction, separation, refining, and alloying.[5] A particular focus of this Executive Order is to ensure a stable supply chain for rare earth elements. The rare earth elements include praseodymium (used as noted in Chap. 3 to produce a bright yellow color in ceramic glazes), neodymium, gadolinium, and dysprosium. These elements are important because they are used to enable many different technologies from energy efficient lighting, to cancer treatment drugs, lasers, and medical imaging.

Many clean energy technologies also rely on rare earth materials. Our ability to convert wind into electricity using a wind turbine happens inside the generator where there is a powerful neodymium-iron-boron magnet. These are among the strongest magnets ever made. A single 2-megawatt (MW) wind turbine, large enough to provide electricity for about 400 homes, contains 25 kg of neodymium. At the Wild Horse Wind Farm near Ellensberg in central Washington there are 149 1.8 MW turbines using a total of almost 4,000 kg of neodymium. Currently, there are more than 60,000 wind turbines in the United States with enough capacity to power more than 27 million American homes,[6] using magnets containing almost 1.5 million kg of neodymium.

An approach to reduce supply chain risks of critical minerals such as the rare earth elements is to use alternative materials that are more abundant and less prone to geopolitical concerns, price volatility, and other disruptions in supply. One example is the work at Northeastern University in Boston where researchers are developing iron-nickel alloys to replace neodymium and dysprosium [4].

Another rare earth element that may be more difficult to replace than neodymium is erbium. This element is crucial in the glass optical fibers that send our light-encoded information around the world. The strength of the light signal slowly fades as it travels mile after mile requiring periodic amplification. Erbium ions embedded every so often along the optical fiber interact with the incoming light to boost the signal.

Butters' Law named after Gerry Butter of Lucent's Optical Networking Division, predicts that the amount of data that can be transmitted using optical fiber is doubling every nine months. The impact is that the cost of networking is rapidly coming down, which is one of the reasons that we have unlimited long distance, data, and text. Materials including erbium have helped enable this rapid growth.

One of the recurring themes associated with many of the materials in this book is that of the fortuitous accident. The synthesis of polyethylene by Eric Fawcett and Reginald Gibson was due to accidentally introduced oxygen impurities that had found their way into the process. Teflon was discovered as the result of an experiment to produce a new coolant for refrigerators and air conditioners. The experiment was unsuccessful in meeting its original goal but led instead to a brand new material, the first to set foot on the Moon, and a multibillion dollar industry. Jan Czochralski's clumsiness became a new technique for growing large crystals that was essential for the mass production of silicon chips. And Charles Goodyear's decade of persistence was eventually rewarded by a process for the stabilization of rubber.

We continue to see serendipity play an important role in our discovery of new materials. One more recent example is the discovery of carbon nanotubes by Sumio Iijima, a Senior Research Fellow at NEC Corporation in Japan [5]. Iijima and his colleagues were studying the molecular structure of fullerenes (another form of carbon) and how to mass produce it. They were also experts in high-resolution electron microscopy, the invaluable instrument to identify nanoscale structures. During one experiment, Iijima noticed in the electron microscope image long very thin needle-like structures. These structures are very different from the spherical fullerenes that were the object of the study. Iijima named these materials carbon nanotubes [6]. The identification of the carbon nanotube led to an enormous amount of research on various forms of carbon—the nanohorn, the nanocone, the nano onion, the nanobelt, which eventually led to the isolation of graphene—one atom thick sheets of carbon Graphene has been proposed for many applications including nanoribbon transistors, which may one day replace the silicon chip at the heart of our electronic world.

A challenge faced by many materials is how to scale-up production to make them commercially viable. The discovery of carbon nanotubes was followed by a host of proposals for their use in exciting and unusual applications, including commercial airplanes that are so light that once airborne they require almost no fuel to an elevator to space. None of these grand technologies have been realized. Actual applications have been much more modest—primarily as a composite in high-end bowling balls and sports equipment.

One of the reasons for the lack of commercial success is that low cost, industry scalable manufacturing processes did not follow the scientific discovery. Graphene is presently finding itself in a very similar position to carbon nanotubes.[7]

This challenge was faced by aluminum, which remained a precious metal from when it was first made in 1825 until 1886—sixty-one years—when Charles Hall and Paul Héroult independently showed that alumina could be dissolved in cryolite and the metal extracted by electrolysis. On the other hand, commercial production of Teflon lagged only 8 years after its discovery. The process was patented by Malcolm Renfrew. In addition to describing its synthesis, the patent states: A further object is to provide such a process which is economical and relatively simple to carry out.[8]

According to a Friends of the Earth report we extract and use around 50% more natural resources than we did only 30 years ago, at about 60 billion tons of raw materials a year.[9] Global consumption is expected to rise to over 100 billion tons by 2030 and possibly more than 180 billion tons by 2050 [8]. For the majority of human history our use of materials has followed a simple path: *take—make—use—dispose* [9]. This is still what we do with the vast majority of the materials we use today. An entire history of plastics from 1950 to 2015 found that almost 79% of the 8,300 million metric tons of plastic produced during this sixty five year period ended up accumulated in landfills or in the natural environment, 12% was incinerated, and only 9% had been recycled. If current production and waste management trends continue as much as 12,000 million metric tons of plastic waste will be in landfills or the natural environment by 2050 [10].

Increasingly we are becoming aware of the environmental impact of our consumption, the impact that it has on climate change as well as the many social and health effects. These concerns are driving the materials community to consider a circular materials economy where we reuse, recondition, and recycle where possible and only then, as a very last resort, consider landfill or combustion.

Advances in materials science and processing technology have allowed the silicon chip to keep pace with the projections of Moore's law. Rose's law is a quantum computing version of Moore's law for semiconductor-based machines and predicts the exponential growth of the computing power of quantum computers. Many of these powerful next generation computers are based on silicon; a new application for the material that revolutionized computers the first time around [11].

We will continue to explore our use of existing materials and to discover new ones. We are in an exciting era for materials science with the rapid development of nanomaterials like carbon nanotubes, exciting two-dimensional materials such as graphene, and superatoms—clusters of a handful of atoms with unexpected and unpredictable properties that can be changed one atom at a time. There are "smart" materials that can remember their shape, repair themselves, or self-assemble into new structures.

Even Hollywood has got into the materials game. In the Walt Disney Pictures movie *Black Panther*, the world order is dominated by a super alloy called vibranium, which was deposited on Earth by a meteorite 10,000 years ago. (The first iron tools were formed from meteoric iron.) Vibranium has many superior properties including an exceptional strength. Because of its meteoric origin vibranium is rare, which makes Wakanda, the country where it was found, one of the most technologically advanced nations. Maybe there is also a new, yet to be discovered, alloy out there that can rival the fictional vibranium.

Notes

1. Moore's Law is that the number of transistors on integrated circuits doubles approximately every two years. The law was described in 1965 by Intel co-founder Gordon Moore.
2. Data is for 2019 from the United States Geological Survey (usgs.gov).
3. *Disclosing the Use of Conflict Minerals*; United States Securities and Exchange Commission: Washington DC, 2013. A detailed publication on conflict minerals is: Fitzpatrick, Colin, Olivetti, Elsa, Miller, T. Reed, Roth, Richard & Kirchain, Randolph (2014) Conflict minerals in the compute sector: Estimating extent of tin, tantalum, tungsten, and gold use in ICT products, *Environ. Sci. Technol.* 49, 974–981.
4. https://appleinsider.com/articles/20/02/06/apple-removed-18-smelters-and-ref iners-in-2019-for-flouting-conflict-mineral-code-of-conduct

5. A Federal Strategy to Ensure Secure and Reliable Supplies of Critical Minerals, U.S. Department of Commerce, June 4, 2019 https://www.commerce.gov/news/reports/2019/06/federal-strategy-ensure-secure-and-reliable-supplies-critical-min erals
6. American Wind Energy Association, Demand drives wind power development to new heights in first quarter of 2018, https://www.awea.org/resources/news/2018/demand-drives-wind-power-development-to-new-height
7. Geim, A.K, & Novoselov, K.S. (Undated). The rise of graphene https://arxiv.org/pdf/cond-mat/0702595.pdf
8. Renfrew, M.M. (1950). Polymerization of tetrafluoroethylene with dibasic acid peroxide catalysts *US Patent 2,534,058*. The patent application was filed in November 1946 and issued in December 1950. Renfrew Hall at the University of Idaho is named in honor of Malcolm Renfrew where he served as head of chemistry. Malcom Renfrew passed away at the age of 103.
9. Overconsumption—Our use of the world's natural resources, Friends of the Earth UK https://arxiv.org/pdf/cond-mat/0702595.pdf

References

1. Autesserre, S. (2012). Dangerous tales: Dominant narratives on the DRC and their unintended consequences. *African Affairs, 111*, 202–222.
2. Kara, S. (2019). I saw the unbearable grief inflicted on families by cobalt mining. I pray for change *The Guardian*. This article appeared on Monday 16 December 2019.
3. Fu, X., Beatty, D. N., Gaustad, G. G., Ceder, G., Roth, R., Kirchain, R. E., Bustamante, M., Babbit, C., & Olivetti, E. A. (2020). Perspectives on cobalt supply through 2030 in the face of changing demand. *Environmental Science and Technology, 54*, 2985–2993.
4. Lewis, L. H. & Barmak, K. (2019). Rare earth-free permanent magnetic material. *U.S. Patent 10,332,661 B2*. The patent was issued June 2019.
5. Iijima, S. (1991). Helical micro-tubules of graphitic carbon. *Nature, 345*, 56–58.
6. Iijima, S., & Ichihashi, T. (1993). Single-shell carbon nanotubes of 1-nm diameter. *Nature, 363*, 603–605
7. Novoselov, K. S., Geim, A. K., Morozov, S. V., Jiang, D., Zhang, Y., Dubonos, S. V., Grigorieva, I. V., & Firsov, A. A. (2004). Electric field effect in atomically thin carbon films. *Science, 306*, 666–669
8. Olivetti, E. A., & Cullen, J. M. (2018). Toward a sustainable materials system. *Science, 360*, 6396.
9. Ashby, M. F. (2016). *Materials and Sustainable Development* (p. 219). Amsterdam: Elsevier.

10. Geyer, R., Jambeck, J. R., & Lavender Law, K. (2017). Production, use, and fate of all plastics ever made. *Science Advances, 3,* e1700782.
11. Maurand, R., Jhl, X., Kotekar-Patil, D., Corna, A., Bohuslavskyi, H., Laviéville, R., Hutin, L., Barraud, S., Vinet, M., Sanquer, M., & De Franceschi, S. (2016). A CMOS silicon spin qubit *Nature Communications, 7,* 13575.

Index

A

Africa 2, 10, 12, 13, 18, 30, 32, 49, 65, 70, 73, 74, 76, 77, 150, 162, 199
Amazonas 125, 126, 133, 134
Anglicus, Bartholomaeus 71
Artificial Intelligence 2, 188
Asia 10, 13, 14, 34, 49, 50, 54, 75, 147
Aspdin, Joseph 107, 108, 111, 112
Atahualpa 71
Attenborough, Sir David 10

B

Bardeen, John 188, 190
Bayer, Karl Joseph 164
Beaker people 15, 16
Bell Laboratories 101, 184, 186–191
Bennett, Arnold 40
Berzelius, Jöns Jacob 181, 187
Bessemer, Sir Henry 55–57
Boeing 114, 161, 170, 171, 173
Bone China 34, 35
Böttger, Johann Friedrich 33, 34, 40

Brattain, Walter 188–190
Bravais, August 99, 100
Brazil 61, 71, 120, 126, 133, 134, 145, 147
British Museum 19, 20, 40, 67, 94, 95, 98
Bronze 1–3, 5, 14, 15, 19, 49, 50, 54, 56, 65
Bronze Age 1, 2, 15, 47, 50, 66
Brown Bess 18
Brunel, Isambard Kingdom 52
Brunel, Sir Marc Isambard 111
Burj Khalifa 61, 122

C

Cancer 4, 82, 83, 200
Candela, Felix 119, 120
Carbon dioxide 31, 49, 51, 81, 92, 120, 121, 164, 182
Carbon monoxide 31, 42, 49, 53, 55, 81, 164
Carnegie, Andrew 57, 58
Catalyst 4, 13, 42, 80, 81, 89, 137, 157

Chihuly, Dale 87, 92
China 2, 15, 18, 31, 32, 34, 35,
    46, 61, 76, 77, 119–122, 145,
    156, 173, 182
Columbus, Christopher 127
Concorde 161, 162, 171
Copper 2, 3, 15, 49, 50, 66–69, 72,
    75, 78, 79, 145, 157, 167,
    168, 172, 187, 188, 198
Cornell University 4
Czochralski, Jan 183, 184, 201

D

Damascus sword 55
Darwin, Charles 10
Dash, William 184
Da Vinci, Leonardo 72
Davy, Sir Humphry 163, 181
Deville, Henri Sainte-Claire 162,
    163, 168, 181, 194
Diamond 5, 9, 36, 65, 77, 98, 104,
    166, 178, 179, 199
Dislocation 53, 167, 168, 172, 174,
    184
Dolní Vestonice 28
Drake, Sir Francis 4, 71
Dunlop, John Boyd 132, 133
DuPont 136, 152

E

Earth's crust 27, 48, 73, 96, 168,
    181, 191, 198
Earthenware 25, 31, 32, 34, 35, 40,
    41
Egypt 36, 47, 70, 77, 119, 134
Eiffel Tower 52
Einstein, Albert 66
Electrolytic Marine Salts Company
    75
England 9, 11, 20, 34, 37, 39–41,
    54, 56, 69, 71, 110, 116, 125,
    129, 146, 148

Europe 11, 12, 15, 18, 19, 34,
    38–40, 49, 50, 54, 71, 75, 95,
    126, 127, 134, 136, 137, 171

F

Fire 2, 16–18, 30, 35, 41, 72, 107,
    108, 116, 122, 127, 130, 163,
    198
Fleming, Sir John Ambrose 185
Flintlock 17, 18, 21
Ford, Henry 133, 134, 171
Frere, John 19
Furnace 30, 31, 42, 43, 49–51, 58,
    60, 61, 89, 90, 92, 182

G

Galilei, Galileo 100
Glaze 35–39, 43, 96–98, 200
Godalming 186
Golem 25
Goodyear, Charles 130, 131, 137,
    141, 142, 165, 201
Grand Coulee Dam 114, 115, 182
Graphene 5, 6, 31, 201–203
Greece 2, 11, 15, 68
Grégoire, Marc 153, 154
Grime's Graves 14, 18, 19

H

Hall, Charles Martin 163
Hancock, Thomas 128–132
Handaxe 2, 3, 11, 12, 16, 18, 19
Haruta, Matsutake 80, 81
Harvard University 92
Hematite 43, 48, 49, 103
Henry VIII 4, 65
Heraeus 79
Héroult, Paul Louis Toussaint 163,
    202
Hittite 47, 55
Hockham, George 101, 104

Homo erectus 2, 11, 12, 18
Homo heidelbergensis 2
Homo sapiens 11
Hooke, Robert 99, 100, 102
Hoover Dam 113–115
Hyatt, John Wesley 150, 151, 158

I

Ice Age 14
Imperial Chemical Industries (ICI)
    146, 149
Imperial College 56, 82
Ingalls Building 118
Intel 190, 192, 197
Internet of Things 2, 188, 193
Iron Age 1, 3, 10, 48, 50
Iron Bridge 45
Isle of Wight 99, 148

J

Jennings, Hamlin 112, 113
Jomon 30
Journal of Materials Science 3

K

Kao, Charles 101
Kaolin (China clay) 27, 33, 34, 42
Karat 68
Kelly, William 56
Kilby, Jack 79, 191, 192
Kursk Magnetic Anomaly 61

L

Lead 27, 36, 37, 42, 43, 50, 60, 72,
    74, 75, 97, 98, 102, 104, 110,
    114, 115, 128, 130, 132, 168,
    191, 194
Leakey, Louis and Mary 10, 11
Liberty ships 60
Limonite 48, 49

Lubbock, John 1

M

Macintosh, Charles 129, 197
Magnetite 48, 49, 61, 96, 103
Manaus 126, 133, 134, 145
Mansa Musa 76, 77
Martensite 54
Matchlock 17, 18
McCourt, Jon 139
Mediterranean 15, 70, 151
Meissen 34, 39, 40
Mesabi Iron Range 49
Mesopotamia 26, 32
Meteorite 47, 203
Midas 75
Middle East 12, 30, 36
Mohs, Fredrich 9, 166
Moore, Gordon 190, 192, 203
More, Sir Thomas 4, 65, 78

N

Nanjing 5, 25, 38
Nanomaterial 4, 203
Nanoparticle 4, 80–83, 95, 96, 157
Nanosprings 5
Nanotechnology 4, 80, 81, 83
Neanderthal 11
Near East 15
Newton, Sir Isaac 69, 72
New York City 7, 47, 52, 58, 59,
    122, 135, 150, 169, 170, 186,
    189
Northern Ireland 139, 140
Northwestern University 3
Noyce, Robert 190–192

O

Obsidian 3, 4, 87, 96, 103
Olduvai Gorge 10, 11

P

Paleolithic 1, 16, 20, 87
Pantheon 108–110, 115
Paracelsus 72, 84
Periodic Table of Elements 46, 66,
     96, 166, 178
Phillips, John George 186
Pliny the Elder 90, 113
Plunkett, Roy 152, 154
Plutonium 5, 114
Polk, James 76
Porcelain 33–35, 38–40, 42
Potteries, The 1, 16, 21, 26, 27,
     30–33, 35, 37–41, 125
Priestley, Joseph 97, 128
Pullman, George M. 165, 174
Purple of Cassius 95, 103

Q

Quantum computer 193, 194, 203
Quartz 3, 8, 9, 16, 20, 27, 33, 34,
     70, 73, 74, 92, 94, 97, 99, 181
Queen Elizabeth I 4

R

Radar 148, 149, 170, 179
Rapid Diagnostic Test (RDT) 82
Ravenscroft, George 97, 98
Recycling 21, 60, 61, 138, 145, 146,
     156, 170, 200
Renfrew, Malcolm 153, 154, 202
River Thames 20
Roman 9, 10, 16, 19, 30, 37, 51, 65,
     68–71, 90, 95, 100, 108–110,
     112, 113, 122, 130
Roosevelt, Franklin D. 69, 137, 138,
     200

S

Sand 32–36, 66, 74, 88–90, 96,
     102, 103, 108, 153, 168, 182,
     183

Schultheis, Michael 108, 109
Sèvres 39, 40
Shakespeare, William 9
Shockley, William 187, 188, 190
Silicon Age 2
Silly Putty 139
Simpson, Edward "Flint Jack" 19, 20
Smeaton, John 110
Spode, Josiah 35
Stone Age 1–3, 10, 11, 13, 14, 18,
     20, 99
Stoneware 31–34, 40
Sydney Opera House 119

T

Teal, Gordon 184
Texas Instruments 79, 191, 192
Thiokol 135, 136
Thomas, Sidney Gilchrist 57, 62
Thomsen, Christian Jürgensen 1
Three Age system 1–3
Three Gorges Dam 115, 120
Tin 2, 3, 15, 36–38, 50, 183, 198,
     199
Tutankhamen 47, 55, 70
Twyford, Thomas 38

U

United States 12, 26, 34, 37, 42,
     56, 58–61, 69, 73, 75, 78, 82,
     101, 113, 114, 116–118, 120,
     122, 133, 135–139, 141, 146,
     162–164, 166, 170, 171, 173,
     174, 182, 194, 199, 200, 203
United States Geological Survey
     (USGS) 61, 62, 78, 84, 157,
     173, 203
Uranium 5, 154

V

Varna 66, 83

Vauxhall 134
Venus figurine 28
Versailles 40, 91, 98
Vibranium 203
Von Tschirnhaus, Count Ehrenfried
    Walther 33, 34, 42
Vulcanization 130, 131, 137, 141,
    142

W

Wedgwood, Josiah 33

Weibull, Waloddi 115
World War I 49, 134, 135, 147, 169
World War II 49, 60, 82, 114, 137,
    147, 148, 154, 169, 170, 179,
    180
Wright Flyer 168, 169
Wright, Frank Lloyd 118

Z

Zachariasen, William Houlder 94

Printed in the United States
by Baker & Taylor Publisher Services